Logic and Logic Design

Logic and Logic Design

B. GIRLING and H. G. MORING

Department of Mathematics
The City University, London

INTERTEXT BOOKS

Published by
International Textbook Company Limited
24 Market Square, Aylesbury, Bucks.

First published 1973

ISBN 0 7002 0201 3

Text set in 10/12 pt. IBM Press Roman, printed by photolithography,
and bound in Great Britain at The Pitman Press, Bath

Contents

v

Chapter 1
Basic Number Systems

Introduction

Although this text is intended to be a basic introduction to Boolean logic and its applications, most of the logic examples will be orientated towards aiding the manipulation of numerical quantities. Therefore, a preliminary discussion of the fundamental number systems involved, and the elementary numerical operations associated with them, will take place before the introduction to sets and Boolean algebra.

With the advent of digital computing machines, and the consequent employment of two-state devices, a number system based upon the radix-2, i.e. binary system, was found to have distinct practical advantages over the conventional radix-10, i.e. decimal or denary system. However, the basic theory of number manipulation which follows is independent of the radix, and three systems will be discussed in detail, namely binary, octal (radix-8), and decimal. A list of seven radices and their names are given later in Table 1.6.

Any number can be represented as a sum of powers of the radix—rather like a polynomial in which the variable is replaced by the radix and all the coefficients are positive and lie between zero and that integer which is one less than the radix. The evolution of this from the abacus or bead frame can easily be shown. Consider the numbers 427 and 305, both in the decimal system. Figure 1.1(a) shows 427 and Figure 1.1(b) shows 305. In both cases, wires for units, tens, hundreds, and thousands are shown. Note that on transferring these numbers to manuscript a 0, or zero symbol, is required for any wire which is free of beads. Non-significant zeros, of course, are not usually shown. For example, in the above illustrations, the numbers are 0427 and 0305 and in both cases the non-significant leading zeros are omitted.

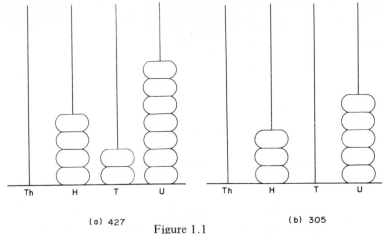

(a) 427

(b) 305

Figure 1.1

When a decimal number is read, it is essentially translated as powers of ten. For example,

$$427 = 4 \times 100 + 2 \times 10 + 7 \times 1$$
$$= 4 \times 10^2 + 2 \times 10^1 + 7 \times 10^0$$

Therefore it can be seen that a decimal integer has the mathematical form of

$$\sum_{i=0}^{n} (A_i \times 10^i) \qquad \begin{cases} n \geqslant 0 \\ A_i = \{0, 1, 2, 3, \ldots, 9\} \end{cases} \dagger$$

preceded by a positive or negative sign, whilst a decimal fraction has that of

$$\sum_{i=0}^{m-1} (A_i \times 10^{i-m}) \qquad \begin{cases} m \geqslant 1 \\ A_i = \{0, 1, 2, \ldots, 9\} \end{cases} \dagger$$

with the appropriate sign attached. Therefore, a mixed number, i.e. one consisting of an integral part and a fractional part separated by a decimal point, has the general form of

$$\sum_{i=0}^{n+m} (A_i \times 10^{i-m}) \qquad \begin{cases} n \geqslant 0 \\ m \geqslant 1 \\ A_i = \{0, 1, 2, \ldots, 9\} \end{cases} \dagger$$

with the appropriate sign attached.

† A_i can take *any* of the values within the brackets.

This approach can be extended to any other radix by replacing the 10, in the above expressions, by the new radix and the set of integers A_i by the set lying between zero and that integer which is one less than the radix in question. For example, octal numbers, i.e. numbers having a radix of eight, can be expressed as

$$\sum_{i=0}^{n+m} (A_i \times 8^{i-m}) \quad \begin{cases} n \geqslant 0 \\ m \geqslant 1 \\ A_i = \{0, 1, 2, \ldots, 7\} \,\dagger \end{cases}$$

and binary numbers (radix-2) as

$$\sum_{i=0}^{n+m} (A_i \times 2^{i-m}) \quad \begin{cases} n \geqslant 0 \\ m \geqslant 1 \\ A_i = \{0, 1\} \,\dagger \end{cases}$$

Decimal, octal, and binary representations of the same quantities are shown in Tables 1.1, 1.2, and 1.3. D, $\emptyset$, and B will be the notation used to differentiate between decimal, octal, and binary respectively, non-significant zeros being omitted.

Table 1.1

D	B	$\emptyset$
0	0	0
1	1	1
2	10	2
3	11	3
4	100	4
5	101	5
6	110	6
7	111	7
8	1000	10
9	1001	11
10	1010	12

Table 1.2

D	B				$\emptyset$	
	1			1		1
	10		1	010		12
	100		1 100	100		144
1 000		1 111	101	000	1	750
10 000	10 011	100	010	000	23	420

Obviously, the decimal notation is the most economical of the three, in terms of the actual number of digits which must be written down for any particular value, whilst the binary notation is the most economical in symbols since only

$\dagger$ A_i can take *any* of the values within the brackets.

two are used. However, the most important practical reason for choosing the binary system is the ease with which it can be used to describe the state of an electronic device, which can be switched either on or off. The octal notation is a sort of half-way house and is used extensively in computer programming. The principal reason for this is that computer *instructions* as well as *numerical values* can be memorized in octal more easily than in binary.

Table 1.3

B	$\emptyset$	D
1	1	1
10	2	2
100	4	4
1 000	10	8
10 000	20	16

Radix Conversion

In this section, it will be convenient to use the notation $(N)_r$ to denote the number N using the radix r. A number expressed in *any* radix can be transformed to any other radix. In particular, the three systems mentioned above will be considered.

(a) DECIMAL INTEGER TO BINARY INTEGER

Divide the decimal integer by 2 and place the remainder in the least significant position, i.e. the right-hand side, of the answer. Repeat, filling the next least significant position, until the quotient becomes 0. This is the indication that the conversion is complete and the last remainder is the most significant digit. For example, convert $(132)_{10}$ to binary.

$$
\begin{array}{lll}
132 \div 2 = 66 & \text{remainder} & 0 \ (\text{least significant digit}) \\
66 \div 2 = 33 & & 0 \\
33 \div 2 = 16 & & 1 \\
16 \div 2 = \ 8 & & 0 \\
8 \div 2 = \ 4 & & 0 \\
4 \div 2 = \ 2 & & 0 \\
2 \div 2 = \ 1 & & 0 \\
1 \div 2 = \ 0 & & 1 \ (\text{most significant digit})
\end{array}
$$

i.e.

$$(132)_{10} = (1\ 0\ 0\ 0\ 0\ 1\ 0\ 0)_2$$

(b) DECIMAL INTEGER TO OCTAL INTEGER

As for (a) above, with two replaced by eight. For example, convert $(215)_{10}$ to octal.

$$215 \div 8 = 26 \text{ remainder } 7 \text{ (least significant digit)}$$
$$26 \div 8 = 3 \qquad\qquad 2$$
$$3 \div 8 = 0 \qquad\qquad 3 \text{ (most significant digit)}$$

i.e.

$$(215)_{10} = (327)_8$$

(c) BINARY INTEGER TO DECIMAL INTEGER

Reading from the least significant end, each binary position is given a *weight*. These weights are the decimal numbers $1, 2, 4, 8, 16, \ldots, 2^r, \ldots$, etc. The existence of a 1 in the binary number means that the corresponding weight is to contribute to the decimal total, the weights corresponding to the 0s being ignored. For example, convert $(1\ 1\ 0\ 0\ 1\ 1\ 1)_2$ to decimal.

Reading from the right, the weights are $1, 2, 4, 8, 16, 32, 64$. The decimal equivalent is therefore

$$64 + 32 + 4 + 2 + 1 = 103$$

i.e.

$$(1\ 1\ 0\ 0\ 1\ 1\ 1)_2 = (103)_{10}$$

(d) BINARY INTEGER TO OCTAL INTEGER

First group the digits into threes from the least significant end, i.e. the right, and add any non-significant zeros necessary to make up an exact number of threes. Then use Table 1.4 to effect the necessary conversion. For example, convert $(1\ 1\ 1\ 0\ 0\ 0\ 0\ 0\ 0\ 1\ 1\ 0\ 1)_2$ to octal. $1\ 1\ 1\ 0\ 0\ 0\ 0\ 0\ 0\ 1\ 1\ 0\ 1$ becomes $0\ 1\ 1\quad 1\ 0\ 0\quad 0\ 0\ 0\quad 0\ 0\ 1\quad 1\ 0\ 1$ after adding a leading zero and the octal representation is seen to be $(34015)_8$.

Table 1.4

φ	0	1	2	3	4	5	6	7
B	000	001	010	011	100	101	110	111

(e) OCTAL INTEGER TO BINARY INTEGER

A short cut is possible here. Each octal digit is replaced by three binary digits from Table 1.4 and leading non-significant digits are dropped. For example, convert $(3071)_8$ to binary.

3071 becomes 0 1 1   0 0 0   1 1 1   0 0 1 which, upon dropping the leading non-significant zero, yields $(1\ 1\ 0\ 0\ 0\ 1\ 1\ 1\ 0\ 0\ 1)_2$ as the result.

(f) OCTAL INTEGER TO DECIMAL INTEGER

As for binary to decimal except that the multipliers are now 8 instead of 2. For example, $(3107)_8$ becomes $((3 \times 8 + 1) \times 8 + 0) \times 8 + 7$ which expands to $(1067)_{10}$. Alternatively the octal number can be converted to binary and the procedure of (e) above carried out.

Binary and Decimal Fraction Systems

The integer representations of the three principal number systems have been discussed in some detail and the emphasis will now be placed upon the binary and decimal fraction systems. It will not always be possible to get an exact conversion from a decimal fraction to a binary fraction, the exceptions being when the decimal fraction is in fact a combination of negative powers of two. Usually, therefore, approximations will have to suffice.

(a) DECIMAL FRACTION TO BINARY FRACTION

Here a doubling technique is used, any carry generated filling in the binary digits from the most significant end. For example, to convert 0·62 in decimal to binary, stopping when the seventh binary digit has been generated, the working is set out as follows:

$0.62 \times 2 = 1.24$ therefore record 0·1 with remainder 0·24
$0.24 \times 2 = 0.48$ therefore attach a 0 with remainder 0·48
$0.48 \times 2 = 0.96$ therefore attach a 0 with remainder 0·96
$0.96 \times 2 = 1.92$ therefore attach a 1 with remainder 0·92
$0.92 \times 2 = 1.84$ therefore attach a 1 with remainder 0·84
$0.84 \times 2 = 1.68$ therefore attach a 1 with remainder 0·68
$0.68 \times 2 = 1.36$ therefore attach a 1 with remainder 0·36.

The resulting binary fraction is therefore 0 · 1 0 0 1 1 1 1. Obviously, the next digit would have been a zero, had the process been continued, so that in this case it can be said that $(0 \cdot 1\ 0\ 0\ 1\ 1\ 1\ 1)_2$ is the binary equivalent of $(0.62)_{10}$ the answer being correct to seven bits. A *bit* is the abbreviation used for *binary digit*. Note that the error involved is less than half a unit in the least significant place, i.e. in the above example the error is less than $\frac{1}{2} \times 2^{-7}$. This is analogous to decimal round-off when a number quoted as being correct to n places of decimals has a possible error of $\frac{1}{2} \times 10^{-n}$.

(b) BINARY FRACTION TO DECIMAL FRACTION

A halving technique is used here. Starting with the least significant digit, halve it, then add the next significant digit to the result, and repeat the procedure. The process terminates when the most significant digit has been added in and the value obtained halved irrespective of whether the most significant digit was a 1 or a 0. For example, convert $(0 \cdot 0 1 0 1 1 0 1)_2$ to decimal. The result is $(((((((1 \div 2) + 0) \div 2 + 1) \div 2 + 1) \div 2 + 0) \div 2 + 1) \div 2 + 0) \div 2$ yielding $(0 \cdot 3515625)_{10}$. Note that this type of conversion is exact.

(c) RAPID CONVERSION OF RECURRING BINARY FRACTIONS TO DECIMAL FRACTIONS

The fraction $(0 \cdot 0011 \quad 0011 \quad 0011 \quad 0011 \quad \ldots \ldots)_2$ can be treated as the sum of an infinite geometric progression, namely

$$S = a + ar + ar^2 + ar^3 + ar^4 + \ldots$$

$$= \frac{a}{1 - r}$$

taking $(0 \cdot 0 0 1 1)_2$ as a, $(0 \cdot 0 0 0 0 \quad 0 0 1 1)_2$ as ar, $(0 \cdot 0000 \quad 0000 \quad 0011)_2$ as ar^2, etc., it can be seen that $r = 2^{-4}$ and $a = 3 \times 2^{-4}$. It is perhaps more convenient to work with vulgar decimal fractions at this point and so a is written as $(3/16)_{10}$ and r as $(1/16)_{10}$ and the required sum becomes

$$\frac{3/16}{1 - 1/16} = \frac{3}{15}$$

$$= \frac{1}{5}$$

i.e.

$$(0 \cdot 0011 \quad 0011 \quad 0011 \quad \ldots)_2 = (0 \cdot 2)_{10}$$

In the above example, the infinite binary fraction was replaced by a finite decimal fraction. This will not always be the case as the next example will demonstrate.

The fraction $(0 \cdot 0111 \quad 0111 \quad 0111 \quad \ldots \ldots)_2$ yields by the above procedure

$$a = (0 \cdot 0111)_2 = \left(\frac{7}{16}\right)_{10}$$

$$r = 2^{-4} = \frac{1}{16}$$

and the sum is therefore

$$\frac{7/16}{1-1/16} = \frac{7}{15}$$

which yields as a recurring fraction $(0.4666 \quad 6666 \quad 6666 \quad \ldots)_{10}$. In both the above examples the repetition of the binary fraction started immediately after the decimal point. This is not a necessary condition for the technique to work as can be seen by the conversion of $(0.11001010101010\ldots)_2$ to decimal. In this type of example the non-repetitive part is first removed, e.g.

$$(0.110010101010\ldots)_2 = (0.110 \quad 01 \quad 01 \quad 01 \quad 01 \quad 01 \quad \ldots\ldots)_2$$
$$= (0.110)_2 + (0.00001 \quad 01 \quad 01 \quad 01 \ldots)_2.$$
$$= \left(\frac{6}{8}\right)_{10} + S$$

where

$$S = (0.00001 \quad 01 \quad 01 \quad 01 \quad 01 \quad \ldots\ldots)_2$$

for which

$$a = (0.00001)_2 = (1/32)_{10}$$

and

$$r = 2^{-2} = 1/4$$

Therefore

$$S = \frac{1/32}{1-1/4}$$
$$= \frac{1}{24}$$

Hence

$$(0.1100101010\ldots\ldots)_2 = \left(\frac{3}{4}\right)_{10} + \left(\frac{1}{24}\right)_{10}$$
$$= \left(\frac{19}{24}\right)_{10}$$
$$= (0.791666666\ldots\ldots)_{10}$$

Table 1.5 lists some common decimal fractions and their binary equivalents.

Table 1.5

D	B
0·1	0·0 0011 0011 0011 ....
0·2	0·0 0110 0110 0110 ....
0·3	0·0 1001 1001 1001 ....
0·4	0·0 1100 1100 1100 ....
0·5	0·1 exactly

(d) THE CONVERSION OF OCTAL FRACTIONS

These can be treated as binary fractions except that the multiplier or divisor is now eight instead of two.

Thus $(0·62)_{10}$ yields

$$0·62 \times 8 = 4·96, \quad \text{i.e. } 0·4$$
$$0·96 \times 8 = 7·88, \quad \text{i.e. } 0·47$$
$$0·88 \times 8 = 7·04, \quad \text{i.e. } 0·477$$
$$0·04 \times 8 = 0·32, \quad \text{i.e. } 0·4770$$

Therefore

$$(0·62)_{10} = (0·477)_8 \text{ correct to three digits}$$

Also $(0·504)_8$ is replaced by $(((4 \div 8) + 0) \div 8 + 5) \div 8$ yielding $(0·6325625)_{10}$. To move from binary fractions to octal and vice versa, use Table 1.4. For example, convert (a) $(0·1011101001)_2$ to octal and (b) $(0·5314)_8$ to binary.

(a)
$$(0·1011101001)_2 = (0·101 \quad 110 \quad 100 \quad 100)_2$$
$$= (0·5644)_8$$

(adding 2 zeros at the least significant end).

(b)
$$(0·5314)_8 = (0·101 \quad 011 \quad 001 \quad 100)_2$$
$$= (0·1010110011)_2$$

Conversion from any Radix p to any Other Radix q

The methods of conversion, between binary, decimal, and octal, just considered are all particular examples of the general technique of conversion from one radix to another.

An integer N can be represented in radix p by

$$a_n p^n + \ldots + a_2 p^2 + a_1 p^1 + a_0 p^0$$

where the a's are chosen from the set of integers 0 to $p - 1$ inclusive. N can also be represented in radix q by

$$b_m q^m + \ldots + b_2 q^2 + b_1 q^1 + b_0 q^0$$

where as before the b's are chosen from the set of integers 0 to $q - 1$ inclusive. Normally, but not necessarily, the integers n and m will be distinct, for example $(13)_{10}$ corresponds to $(15)_8$, or $(1101)_2$, since

$$1 \times 10^1 + 3 \times 10^0 = 1 \times 8^1 + 5 \times 8^0$$
$$= 1 \times 2^3 + 1 \times 2^2 + 0 \times 2^1 + 1 \times 2^0$$

The remainder theorem, upon which integer conversion depends, can be simply stated as follows.

'Given two positive integers N and q, there exists two unique non-negative integers Q and r such that,

$$\text{for } 0 < r < q, \quad \frac{N}{q} = Q + \frac{r}{q}\text{'}$$

Therefore, since

$$\frac{N}{q} = b_m q^{m-1} + \ldots + b_2 q^1 + b_1 q^0 + \frac{b_0}{q}$$

it follows that

$$Q = b_m q^{m-1} + \ldots + b_2 q^1 + b_1 q^0$$

and

$$r = b_0$$

Further, using the same technique

$$\frac{Q}{q} = Q' + \frac{r'}{q}$$

which in this case yields

$$Q' = b_m q^{m-2} + \ldots + b_2 q^0$$

and

$$r' = b_1$$

The procedure to be adopted is as follows.

Divide the given integer, radix p, by q, the required radix; the *remainder is the least significant digit* of the new representation and the *quotient* will be used to determine the remaining digits. The above procedure is then repeated, replacing the given integer by the quotient, just obtained, at each step until the quotient itself is less than the new radix. This last quotient will be the most significant digit in the new representation.

The divisions above must all be carried out bearing in mind the fact that N and the quotients will all be expressed in radix p. For example, to convert $N = (15)_8$ to binary, the working is

$$(15)_8 = 1 \times 8^1 + 5 \times 8^0$$

and therefore

$$\frac{1 \times 8^1 + 5 \times 8^0}{2} = 6 \times 8^0 + \frac{1}{2}$$

thus

$$N = \ldots + 1 \times 2^0$$

$$\frac{6 \times 8^0}{2} = 3 \times 8^0 + \frac{0}{2}$$

thus

$$N = \ldots + 0 \times 2^1 + 1 \times 2^0$$

$$\frac{3 \times 8^0}{2} = 1 \times 8^0 + \frac{1}{2}$$

thus

$$N = \ldots + 1 \times 2^2 + 0 \times 2^1 + 1 \times 2^0$$

and finally

$$\frac{1 \times 8^0}{2} = 0 + \frac{1}{2}$$

thus

$$N = 0 \times 2^4 + 1 \times 2^3 + 1 \times 2^2 + 0 \times 2^1 + 1 \times 2^0$$

or $(1101)_2$ after dropping the non-significant zero at the beginning.

The conversion of fractions reverses the procedure in the sense that multiplication replaces division. A fraction N can be represented in the two systems by

$$a_{-1}p^{-1} + a_{-2}p^{-2} + \ldots + a_{-n}p^{-n}$$

and

$$b_{-1}q^{-1} + b_{-2}q^{-2} + \ldots + b_{-m}q^{-m}$$

with the same restrictions upon the a's and the b's as before. Analogous to the remainder theorem, the relationship $qN = r + P$ holds where q and r are integers and N and P are fractions, in this case in radix p. Therefore since

$$qN = b_{-1} + b_{-2}q^{-1} + \ldots + b_{-m}q^{1-m}$$

it follows that

$$r = b_{-1}$$

and

$$P = b_{-2}q^{-1} + \ldots + b_{-m}q^{1-m}$$

This can be repeated using

$$aP = b_{-2} + b_{-3}q^{-1} + \ldots + b_{-m}q^{2-m}$$

etc.

The procedure to be adopted, in converting fractions, is as follows.

Multiply the given fraction, radix p, by q, the required radix; the *integral part of the product* is the *most significant digit* of the new representation. The fractional part will be used to determine the remaining digits. The above procedure is repeated until the fractional part of the product is zero or *until enough digits have been computed*, which ever occurs first. As before it must be borne in mind that the original and all the intermediate fractions are held radix p.

For example, to convert $(0{\cdot}47)_8$ to binary, the working is as follows.

$$(0{\cdot}47)_8 = 4 \times 8^{-1} + 7 \times 8^{-2}$$

and therefore

$$(4 \times 8^{-1} + 7 \times 8^{-2}) \times 2 = 1 \times 8^0 + 1 \times 8^{-1} + 6 \times 8^{-2}$$

thus

$$N = 1 \times 2^{-1} \ldots$$

$$(1 \times 8^{-1} + 6 \times 8^{-2}) \times 2 = 0 \times 8^0 + 3 \times 8^{-1} + 4 \times 8^{-2}$$

thus

$$N = 1 \times 2^{-1} + 0 \times 2^{-2} \ldots$$

$$(3 \times 8^{-1} + 4 \times 8^{-2}) \times 2 = 0 \times 8^0 + 7 \times 8^{-1}$$

thus

$$N = 1 \times 2^{-1} + 0 \times 2^{-2} + 0 \times 2^{-3} \ldots$$

$$(7 \times 8^{-1}) \times 2 = 1 \times 8^0 + 6 \times 8^{-1}$$

thus

$$N = 1 \times 2^{-1} + 0 \times 2^{-2} + 0 \times 2^{-3} + 1 \times 2^{-4} \ldots$$

$$(6 \times 8^{-1}) \times 2 = 1 \times 8^0 + 4 \times 8^{-1}$$

thus

$$N = 1 \times 2^{-1} + 0 \times 2^{-2} + 0 \times 2^{-3} + 1 \times 2^{-4} + 1 \times 2^{-5} \ldots$$

and finally

$$(4 \times 8^{-1}) \times 2 = 1 \times 8^0 + 0 \times 8^{-1}$$

thus

$$N = 1 \times 2^{-1} + 0 \times 2^{-2} + 0 \times 2^{-3} + 1 \times 2^{-4} + 1 \times 2^{-5} + 1 \times 2^{-6}$$

or

$$(0 \cdot 1\,0\,0\,1\,1\,1)_2$$

The conversion of a mixed number, i.e. one involving an integer and a fraction, should be handled as two separate conversions. The integer part is converted using the remainder theorem and the fractional part using the product technique.

The sexadecimal (hexadecimal) number system (radix-16) is the next stage up from the octal system, i.e. it requires the decimal digits 0 to 9 *plus* six additional, symbols—usually A to F—for its representation. The binary equivalents of the sexadecimal digits consist of four bit groups known as *bytes*.

Other radices occasionally met with, but not discussed further here, are given in Table 1.6.

Table 1.6

Radix	Number system
2	binary
3	ternary
4	quaternary
5	quinary
8	octal
10	{decimal {denary
12	duodecimal
16˙	{sexadecimal {hexadecimal

The Fundamental Arithmetic Operations

The fundamental operations to be considered are those of addition, subtraction, multiplication, and division, augmented here by that of square root extraction. In the following discussion, all the operands will be assumed to be positive.

BINARY ADDITION

The table for binary addition is Table 1.7 where the symbol $\underline{0}$ indicates that there is a 'carry' over into the next most significant bit of the addend. This

Table 1.7

		addend	
+		0	1
augend	0	0	1
	1	1	$\underline{0}$

carry exists if and only if *both* the digits being added are 1s. For example, add 1011 and 1101 in binary (their decimal equivalents being 11 and 26 respectively).

B	D	
1011	11	(augend)
11010	26	(addend)
100101	37	(result)

BINARY SUBTRACTION

The table for binary subtraction is Table 1.8 where the $\underline{1}$ now indicates that a borrow was required from the next most significant bit of the minuend. As an

Table 1.8

		subtrahend	
—		0	1
minuend	0	0	1
	1	1	0

example of direct subtraction consider $(1011011 \cdot 101)_2 - (10001 \cdot 11)_2$, equivalent to $(91 \cdot 625)_{10} - (17 \cdot 75)_{10}$.

B	D	
1011011·101	91·625	(minuend)
10001·11	17·75	(subtrahend)
1001001·111	73·875	(result)

As an alternative to normal binary subtraction both the minuend and the subtrahend can first be converted to octal and subtraction carried out with the aid of an octal subtraction table.

In the above example, the minuend $(1011011 \cdot 101)_2$ is first written as the number $(133 \cdot 5)_8$ and the subtrahend $(10001 \cdot 11)_2$ is written as $(21 \cdot 6)_8$. The octal subtraction thus becomes $(133 \cdot 5)_8 - (21 \cdot 6)_8$, using Table 1.9 where, for example, $\underline{5}$ indicates that the difference is 5 when 1 has been borrowed from the next most significant octal digit of the minuend. Thus

∅		
133·5	(minuend)
− 21·6	(subtrahend)
111·7	(result)

This must then be re-converted to binary in the usual fashion, namely $(111 \cdot 7)_8$ equals $(001 \quad 001 \quad 001 \cdot 111)_2$ which becomes, on neglecting the non-significant zeros, $(1001001 \cdot 111)_2$. This of course checks with the value already found.

Table 1.9

	subtrahend						
−	1	2	3	4	5	6	7
0	7	6	5	4	3	2	1
1	0	7	6	5	4	3	2
2	1	0	7	6	5	4	3
3	2	1	0	7	6	5	4
4	3	2	1	0	7	6	5
5	4	3	2	1	0	7	6
6	5	4	3	2	1	0	7
7	6	5	4	3	2	1	0

BINARY MULTIPLICATION

The table for binary multiplication is Table 1.10, from which it can be seen that the possibility of a carry does not exist with binary multiplication.

Table 1.10

x	0	1
0	0	0
1	0	1

Multiplication is carried out in much the same way as decimal long multiplication except that in the final summation the carry arising from any one column can extend over a number of columns in the sum. For this reason it is advisable to add the rows in pairs ignoring any row consisting only of zeros. The location of the binary point follows closely the procedure for finding the position of the decimal point in multiplication, namely the number of places after the point in the product is equal to the sum of the numbers of places after the points in each of its constituents.

Example. Multiply $(1011 \cdot 1101)_2$ by $(10 \cdot 11)_2$.

$$
\begin{array}{r}
1011 \cdot 1101 \\
\times \quad 10 \cdot 11 \\
\hline
10111 \cdot 101 \\
0000 \cdot 0000 \\
101 \cdot 11101 \\
10 \cdot 111101 \\
\hline
100000 \cdot 011111
\end{array}
$$

The six bits after the binary point check with four bits in the multiplicand and two in the multiplier. The addition of the three significant rows could therefore have been laid out as follows:

$$
\begin{array}{rl}
10111101 & \text{ignoring the binary point} \\
+ \quad 10111101 & \text{ignoring the zero row} \\
\hline
1110110001 & \\
+ \quad 10111101 & \\
\hline
100000011111 &
\end{array}
$$

At this stage the position of the binary point is calculated from the rule given above and the answer is quoted as $(100000 \cdot 011111)_2$ exactly.

Since the multiplier was only correct to two binary places (2B), it may be deemed to be desirable in some instances to round off the answer to the same number of binary places, in which case it is quoted as $100000 \cdot 10$ correct to 2B; the brackets are unnecessary, in this case, since the B indicates that the radix must be 2. Multiplication by 2^n is effected by shifting the binary point n places to the right just as multiplication by 10 in decimal is effected by shifting the decimal point n places to the right. The reverse procedure, i.e. shifting the binary point n places to the left, results in division by 2^n.

BINARY DIVISION

This is carried out by repeated subtraction. It is easier to follow if the divisor is made into a binary integer by first multiplying it by a suitable power of two, in which case the dividend must also be multiplied by the same power of two. For example, treat

$$(1001011 \cdot 1)_2 \div (1 \cdot 0001)_2 \quad \text{as} \quad (10010111000 \cdot)_2 \div (10001 \cdot)_2$$
$$(10111 \cdot 0101001001)_2 \div (0 \cdot 000101)_2 \quad \text{as} \quad (10111010100 \cdot 1001)_2 \div (101 \cdot)_2$$

and

$$(10000111001111 \cdot 01001)_2 \div (1000110000 \cdot)_2 \quad \text{as} \quad (10000111100 \cdot 111101001)_2$$
$$\div (100011 \cdot)_2$$

Example. Divide $(1011101 \cdot 1)_2$ by $(0 \cdot 0101)_2$, giving the answer correct to two binary places.

To make the divisor an integer, shift the binary point four places to the right yielding 101. This must also be done to the dividend yielding 10111010100. In order that the answer be correct to 2B the dividend must have at least three binary digits after the binary point and, since this is not the case, insert three zeros as shown below.

The problem now becomes $(10111010100 \cdot 000)_2 \div (101 \cdot)_2$

```
              100101010·011
101· ) 10111010100·000
          101                    1
           01
            0                    0
           11
            0                    0
          110
          101                    1
           11
            0                    0
          110
          101                    1
           11
            0                    0
          110
          101                    1
           10
            0                    0
          100
            0                    0
         1000
          101                    1
          110
          101                    1
            1
```

The working can obviously be contracted, as is shown below, since it is not necessary to insert a line whenever the multiplier is zero.

```
              100101010·011
101· ) 10111010100·000
          101
           0110
           101
            110
            101
           1000
            101
             110
             101
               1
```

The quotient here has been calculated to 3B and to obtain 2B it is necessary to round off thus

$$(1011101 \cdot 1)_2 \div (0 \cdot 0101)_2 = 100101010 \cdot 10$$

correct to 2B.

BINARY SQUARE ROOTS

The technique for extracting the square root of a binary number follows the decimal pattern very closely. It is, however, somewhat simpler to operate as all the trial divisions are either by 1 or 0. As is the case with decimal square roots, the digits are paired off in both directions from the binary point. If there are an odd number of bits then a zero is added. It must also be remembered that when the last bit of a number is doubled a carry will occur *if it is a 1*.

Example. Evaluate $\sqrt{(1011 \cdot 011)_2}$ quoting the result to 2B. Pairing off in both directions from the binary point and adding non-significant zeros, to enable the result to be calculated to 3B,

$$1011 \cdot 011 \text{ becomes } 10 \quad 11 \quad \cdot \quad 01 \quad 10 \quad 00$$

After each subtraction the next two digits are brought down, the last digit of the previous divisor is doubled and the resulting number is moved one place to the left. The first two steps of the process for this example are as follows:

```
             1    ?
     1  | 10  11  ·  01  10  00
        |  1
   10?  |  1  11
```

```
             1    1      ?
     1  | 10  11  ·  01  10  00
        |  1
   101  |  1  11
        |  1  01
   110? |      10      01
```

With practice the lines corresponding to a zero multiplier can be omitted but for this example they will be included. The example continues with the last three steps.

```
              1   1  ·  0   ?
     1 | 10  11  ·  01  10  00
         1
   101 |  1  11
          1  01
  1100 |     10     01
                     0
 1100? |     10     01  10
```

```
              1   1  ·  0   1   ?
     1 | 10  11  ·  01  10  00
         1
   101 |  1  11
          1  01
  1100 |     10     01
                     0
 11001 |     10     01  10
              1     10  01
11010? |            11  01  00
```

```
              1   1  ·  0   1   0
     1 | 10  11  ·  01  10  00
         1
   101 |  1  11
          1  01
  1100 |     10     01
                     0
 11001 |     10     01  10
              1     10  01
110100 |            11  01  00
                            0
```

, etc.

The square root has been computed to 3B and rounding off gives 11·01 correct to 2B.

Round-off

With binary and decimal integers there are approximately $3\frac{1}{3}$ binary digits for every decimal digit. This is not the case with fractions, for the simple reason that elementary decimal fractions such as 0·1, 0·2, 0·3, etc., can only be represented by a recurring infinite binary decimal as is shown in Table 1.5. It can be seen that 0·2 and 0·4 have the same bit patterns as 0·1 shifted to the left by one and two places respectively. This shows the ease with which binary numbers can be divided or multiplied by powers of two.

To round off a binary fraction to n bits it is only necessary to add a 1 in the $(n + 1)$th position and then ignore the $(n + 1)$th and any further bits. Thus rounding off the binary fraction 0·011101110111 to 10 bits will yield

$$0·01110111011$$
$$\underline{1}$$
$$0·01110111100$$

giving

$$0·0111011110$$

whereas rounding off the same binary fraction to 7 bits will yield

$$0·01110111$$
$$\underline{1}$$
$$0·01111000$$

giving

$$0·0111100$$

The point to notice here is that the process of rounding off can affect several digits and not just the last one retained. If the fractions $(0·2)_{10}$ and $(0·3)_{10}$ (or $(0·1)_{10}$ and $(0·4))_{10}$ from Table 1.5 are added the result is $(0·0111 \quad 1111 \quad 1111 \quad 1111 \quad \ldots \quad \ldots)_2$ whereas, by definition, the value of $(0·5)_{10}$ is $(0·1)_2$. This result is, however, quite in order since $(0·0111 \quad 1111 \quad 1111 \quad 1111 \quad \ldots)_2$ is equal to $(0·1)_2$ correct to any *finite* number of bits, i.e. given an $\epsilon > 0$, it is always possible to find an integer $n > 0$, such that the error in taking $(0·1)_2$ as an approximation to $(0·0111 \quad 1111 \quad 1111 \quad \ldots)_2$ (n digits) is less than ϵ no matter how small ϵ was chosen to be.

The bit patterns for the decimals 0·6 to 0·9 inclusive are simply those for 0·1 to 0·4 inclusive with the first zero after the decimal point replaced by a 1. This is due to the fact that $(0·6)_{10} = (0·1)_{10} + (0·5)_{10}$, etc.

Other Systems

Many variations are possible for representing decimal digits by coded sequences of 0s and 1s. At least four bits are required for the complete range from 0 to 9 inclusive and some codes utilize as many as seven. With a four-bit representation, six of the possible combinations will never occur. With a five-bit representation, this number goes up to twenty-two and with seven there are one-hundred and nineteen redundant combinations.

With some codes, each bit has a particular numerical value attached to it. This value is known as the weight of the bit and the code is said to be a weighted code.

The most common weighted code is natural binary coded decimal (NBCD) and the NBCD code for the ten decimal digits of 0 to 9 inclusive are shown in Table 1.11. The least significant bit has the weight 1, the next weight 2, the

Table 1.11

D	NBCD
0	0000
1	0001
2	0010
3	0011
4	0100
5	0101
6	0110
7	0111
8	1000
9	1001

next weight 4, and the last weight 8. The six combinations for 10 to 15 inclusive do not have any valid meaning in this representation and a number such as 13 (decimal) would be represented in NBCD by 0001   0011. The NBCD code is often referred to by its weights, namely as the 8421 code. *All* the positively weighted four-bit codes are listed in Table 1.12. The ones marked

Table 1.12

8	4	2	1	
7	4	2	1	
7	3	2	1	*
6	4	2	1	
6	3	2	1	
6	3	1	1	*
6	2	2	1	
5	4	2	1	
5	3	2	1	
5	3	1	1	
5	2	2	1	*
5	2	1	1	
4	4	2	1	
4	3	2	1	*
4	3	1	1	
4	2	2	1	
3	3	2	1	

with an asterisk are written out in full in Table 1.13. In both the 6311 and the 5221 codes there are, of necessity, alternative representations of the same decimal number due to the equal weights given to two of the bits. With the 6311 code the decimal digits 1, 4, and 7 can all be represented in two ways and with the 5221 code 2, 3, 7, and 8 can all be represented in two ways.

Table 1.13

D	7	3	2	1	6	3	1	1	5	2	2	1	4	3	2	1
0	0	0	0	0	0	0	0	0	0	0	0	0	0	0	0	0
1	0	0	0	1	0	0	0	1	0	0	0	1	0	0	0	1
2	0	0	1	0	0	0	1	1	0	0	1	0	0	0	1	0
3	0	1	0	0	0	1	0	0	0	0	1	1	0	1	0	0
4	0	1	0	1	0	1	0	1	0	1	1	0	1	0	0	0
5	0	1	1	0	0	1	1	1	1	0	0	0	1	0	0	1
6	0	1	1	1	1	0	0	0	1	0	0	1	1	0	1	0
7	1	0	0	0	1	0	0	1	1	0	1	0	1	1	0	0
8	1	0	0	1	1	0	1	1	1	0	1	1	1	1	0	1
9	1	0	1	0	1	1	0	0	1	1	1	0	1	1	1	0

So far only positively weighted codes have been mentioned but there is no reason why some weights should not be negative as for example in Table 1.14. Again, those marked with an asterisk are written out in full in Table 1.15.

Table 1.14

8	4	(−2)	(−1)
6	2	1	(−2)*
7	3	2	(−1)*
6	2	2	(−1)
5	3	1	(−1)
5	2	2	(−1)

Table 1.15

D	6	2	1	(−2)	7	3	2	(−1)
0	0	1	0	1	0	0	0	0
1	0	0	1	0	0	0	1	1
2	0	1	0	0	0	0	1	0
3	0	1	1	0	0	1	0	0
4	1	0	0	1	0	1	1	1
5	1	0	1	1	0	1	1	0
6	1	0	0	0	1	0	0	1
7	1	0	1	0	1	0	0	0
8	1	1	0	0	1	0	1	1
9	1	1	1	0	1	1	0	1

Many codes can be devised using more than four bits and two such codes are shown in full in Table 1.16.

Table 1.16

D	7 4 2 1 0	5 4 3 2 1 0
0	0 0 0 0 1	0 0 0 0 0 1
1	0 0 0 1 0	0 0 0 0 1 0
2	0 0 1 0 0	0 0 0 1 0 0
3	0 0 1 1 1	0 0 1 0 0 0
4	0 1 0 0 0	0 1 0 0 0 0
5	0 1 0 1 1	1 0 0 0 0 1
6	0 1 1 0 1	1 0 0 0 1 0
7	1 0 0 0 0	1 0 0 1 0 0
8	1 0 0 1 1	1 0 1 0 0 0
9	1 0 1 0 1	1 1 0 0 0 0

Parity Checking

An interesting phenomenon arises whenever there is a bit having a zero weight. Since it is immaterial whether this bit is present or not, as far as the numerical value goes, it can be used for what is known as parity checking. The 74210 code shown above is an example of odd parity. This simply means that each decimal digit has an odd number of bits in its representation, in this case one or three. Table 1.17 shows the even parity version of the 74210 code.

Table 1.17

D	7 4 2 1 0
0	0 0 0 0 0
1	0 0 0 1 1
2	0 0 1 0 1
3	0 0 1 1 0
4	0 1 0 0 1
5	0 1 0 1 0
6	0 1 1 0 0
7	1 0 0 0 1
8	1 0 0 1 0
9	1 0 1 0 0

The usefulness of parity checking codes lies in the fact that logic circuits can be designed to check that the parity is correct. These codes fall into two groups, those which enable errors to be detected and those which enable errors to be both detected *and* corrected. These soon become quite complex and will not be discussed further here.

Other simple codes which are of interest do not have any weights attached to their bits and simply rely upon the patterns involved. For example, a 2-out-of-5 code can be chosen such that for every decimal digit there are precisely two 1s and three 0s as in Table 1.18.

Table 1.18

D	2-out-of-5
0	0 0 0 1 1
1	0 0 1 0 1
2	0 1 0 0 1
3	1 0 0 0 1
4	0 0 1 1 0
5	0 1 0 1 0
6	1 0 0 1 0
7	0 1 1 0 0
8	1 0 1 0 0
9	1 1 0 0 0

Three codes which must be mentioned to complete the picture are the XS3 (excess-three) code, the Johnson code, and the Gray code. These are shown in Table 1.19. The XS3 code is used in what is known as complement arithmetic, a discussion of which follows later. The Johnson and Gray codes rely upon bit patterns of the type known as cyclic, i.e. the representation of each decimal digit differs from its adjacent digits by having exactly one bit different. This makes them exceedingly useful in equipment used for digitizing analog quantities, as a mis-read bit can only cause an error of unity, whereas with NBCD a mis-read bit could cause an error of eight—almost an order of magnitude. Note that in these cyclic codes 0 and 9 are deemed to be adjacent.

Table 1.19

D	XS3	Gray	Johnson
0	0 0 1 1	0 0 0 0	0 0 0 0 0
1	0 1 0 0	0 0 0 1	0 0 0 0 1
2	0 1 0 1	0 0 1 1	0 0 0 1 1
3	0 1 1 0	0 0 1 0	0 0 1 1 1
4	0 1 1 1	0 1 1 0	0 1 1 1 1
5	1 0 0 0	0 1 1 1	1 1 1 1 1
6	1 0 0 1	0 1 0 1	1 1 1 1 0
7	1 0 1 0	0 1 0 0	1 1 1 0 0
8	1 0 1 1	1 1 0 0	1 1 0 0 0
9	1 1 0 0	1 0 0 0	1 0 0 0 0

Sign and Magnitude Arithmetic

In this section it will be convenient to consider adding and subtracting integers capable of being represented by five bits in plain binary representation and *not* in a coded form.

Each integer will be held with a scale factor of 2^{-5} so that $(+13)_{10}$ will be represented by $0 \cdot 01101$ and $(+7)_{10}$ by $0 \cdot 00111$, etc. The zero in the integer position, i.e. to the left of the 'point', indicates that the number is positive. Negative numbers will have a 1 in their integer position so that $(-13)_{10}$ will be represented by $1 \cdot 01101$ and $(-7)_{10}$ by $1 \cdot 00111$.

Thus, in sign and magnitude representation, the integer part of the binary form represents the sign, 0 for positive, and 1 for negative, and the fractional part the magnitude.

ADDITION OF TWO POSITIVE NUMBERS

The only trouble which can arise, in this case, is when the sum is greater than or equal to unity, since an integer part of 1 represents a negative number. For example $1 \cdot 00000$ in sign and magnitude represents $(-0)_{10}$.

A carry over into the sign position from the fractional magnitude part indicates overflow, i.e. the number has exceeded the allowable maximum, and is detected as such in modern computers.

ADDITION OF TWO NEGATIVE NUMBERS

Only the magnitudes are added and the negative sign bit is then attached to the sum. As before, overflow can take place.

ADDITION OF TWO UNLIKE NUMBERS USING 1'S COMPLEMENT

In the sum $a + b = c$, c is referred to as the result, a as the augend, and b as the addend. The 1's complement of the magnitude of a number is obtained by interchanging the 1s and the 0s. For example the 1's complement of $\cdot 01101$ is $\cdot 10010$.

To add two numbers of opposite sign, take the magnitude of the augend and add to it the 1's complement of the addend. If the result overflows then *add 1 to the least significant end of the result* and attach the *sign of the augend*. If the result does not overflow, then form the 1's complement of the result and attach the *sign of the addend*.

The above addition of 1 to the least significant end of the result is known as 'end-around-carry'.

Example. Add $1 \cdot 10110$ and $0 \cdot 10011$, i.e.

$$(-22)_{10} \quad \text{and} \quad (+19)_{10}$$

(a) $1 \cdot 10110 + 0 \cdot 10011$

$\cdot 10110$	augend
$\cdot 01100$	1's complement of addend
$1 \cdot 00010$	intermediate result with carry

$\cdot 00010$	
1	add end-around-carry
$\cdot 00011$	magnitude of result

The sum is $1 \cdot 00011$, or $(-3)_{10}$, since the sign of the augend is 1.

(b) $0 \cdot 10011 + 1 \cdot 10110$

$\cdot 10011$	augend
$\cdot 01001$	1's complement of addend
$\cdot 11100$	intermediate result without carry

The sum is $1 \cdot 00011$, or $(-3)_{10}$, since the sign of the addend in this case is 1.

1's Complement Arithmetic

In sign and magnitude representation of decimal numbers $(+7)_{10}$ is $0 \cdot 00111$ and $(-7)_{10}$ is $1 \cdot 00111$. In the 1's complement representation $(+7)_{10}$ is $0 \cdot 00111$ and $(-7)_{10}$ is $1 \cdot 11000$ where, as before, the 1 in the integer position indicates that the number is negative.

The rule for addition, using this type of number representation, is to add the two numbers *including the sign bits.* If there is an end-around-carry from the sign bit position, then it is added to the least significant bit of the result. Should both numbers be positive, then carry into the sign bit position constitutes overflow. Should both numbers be negative, then overflow is indicated if the digit in the *unit* position is a 0 since the sum cannot be positive.

Example 1.

	$1 \cdot 01101$	augend $(-18)_{10}$
	$1 \cdot 11011$	addend $(-4)_{10}$
	$11 \cdot 01000$	i.e. with end-around-carry
Hence	$1 \cdot 01000$	
	1	
	$1 \cdot 01001$	

The sum is therefore $1 \cdot 01001$ or $(-22)_{10}$.

Example 2.	$0 \cdot 01011$	augend $(+11)_{10}$
	$1 \cdot 10001$	addend $(-14)_{10}$
	$1 \cdot 11100$	i.e. no end-around-carry

Hence, the sum is $1 \cdot 11100$ or $(-3)_{10}$.

Example 3.	$1 \cdot 10001$	augend (-14 in decimal)
	$0 \cdot 11011$	addend ($+27$ in decimal)
	$10 \cdot 01100$	i.e. with end-around-carry

Hence $0 \cdot 01100$
          $\underline{\qquad 1}$
          $0 \cdot 01101$

The sum is therefore $0 \cdot 01101$ or $(+13)_{10}$.

Example 4.	$1 \cdot 10001$	augend $(-14)_{10}$
	$1 \cdot 11010$	addend $(-5)_{10}$
	$10 \cdot 01011$	i.e. with end-around-carry *but* positive sign

Hence overflow has taken place and thus the sum is in fact out of range.

2's Complement Arithmetic

Two methods exist for finding the 2's complement of a number.

Either (a) form the 1's complement and then add a 1 to the least significant bit or (b) start at the least significant end and examine the bits in turn. For each 0 in the original replace it by a 0 in the complement until the first 1 is reached. Copy the first 1 in the original, into the complement and thereafter interchange 1s and 0s for the rest of the number, *including the sign bit.*

Example 1. Form the 2's complement of $0 \cdot 01101$.

(a) The 1's complement of $0 \cdot 01101$ is $1 \cdot 10010$, and

$$1 \cdot 10010$$
$$\underline{\qquad 1}$$
$$1 \cdot 10011$$

Hence, by the first method, the 2's complement of $0 \cdot 01101$ is $1 \cdot 10011$.

(b) Original is $0 \cdot 01101$, 2's complement is $1 \cdot 10011$, where the least significant 1 was preserved and all other 1s and 0s interchanged.

Example 2. Write down the 2's complement of (a) $0 \cdot 01100$ and (b) $1 \cdot 10110$.

(a) The complement is $1 \cdot 10100$.

(b) The complement is $0 \cdot 01010$.

Addition using 2's complements is effected by adding the two numbers *including their sign bits* and neglecting any carry from the sign bit position.

Example 1. Add 1·11011 and 0·01100.

$$
\begin{array}{ll}
1 \cdot 11011 & \text{augend } (-5)_{10} \\
\underline{0 \cdot 01100} & \text{addend } (+12)_{10} \\
10 \cdot 00111 &
\end{array}
$$

The result is 0·00111 or $(+7)_{10}$.

The procedure is the same if both numbers are negative.

Example 2. Add 1·10011 and 1·10101.

$$
\begin{array}{ll}
1 \cdot 10011 & \text{augend } (-13)_{10} \\
\underline{1 \cdot 10101} & \text{addend } (-11)_{10} \\
11 \cdot 01000 &
\end{array}
$$

The result is 1·01000 or $(-24)_{10}$. Overflow is detected when the sign bit is *altered*.

Example 3. 0·11011 + 0·10101

$$
\begin{array}{ll}
0 \cdot 11011 & \text{augend } (+27)_{10} \\
\underline{0 \cdot 10101} & \text{addend } (+21)_{10} \\
1 \cdot 10000 &
\end{array}
$$

the 1 in the sign position indicating overflow.

With negative numbers this would have become a 0 instead of the necessary 1.

Example 4. By converting 1·10100 − ·11000 to a summation, show that the result overflows.

$$
\begin{array}{ll}
1 \cdot 10100 & \text{augend } (-10)_{10} \\
\underline{1 \cdot 00110} & \text{addend } (-24)_{10} \\
10 \cdot 11010 &
\end{array}
$$

where the 0 in the sign position indicates overflow.

Suggestions for Further Reading

Flores, I., 1963, *The Logic of Computer Arithmetic*, Prentice-Hall, Englewood Cliffs, N. J.

Sudweeks, R. W., 1968, *Digital Techniques*, Pitman, London.

Ware, W. H., 1963, *Digital Computer Technology and Design*, Vol. 1, Wiley, New York.

Exercises

1. Carry out the following integer conversions:
   (a) $(421)_{10}$ to binary and octal
   (b) $(101100101)_2$ to octal and decimal
   (c) $(4075)_8$ to binary, quinary, and decimal

2. Convert the following *exact* fractions to the specified radix:
   (a) $(0.6095)_{10}$ to binary
   (b) $(0.1011011)_2$ to decimal
   (c) $(0.40713)_8$ to decimal

3. Express π and e in binary correct to 10 bits.

$$(\pi = 3.14159265, \quad e = 2.71828183)$$

4. Convert the recurring decimal

$$13.68484848 \ldots$$

   to binary.

5. Convert the recurring binary fraction

$$0.0100011101110111 \ldots$$

   (a) to decimal
   (b) to hexadecimal

6. Find the binary functions representing

$$\frac{1}{n} \text{ for } n = 3, 4, \ldots, 10, 11$$

7. Write down the recurring binary fractions for $(0.1)_{10}$, $(0.02)_{10}$, and $(0.005)_{10}$ and show that their sum equals $(0.001)_2$.

8. Evaluate $A + B$ and $A \div B$:
   (a) when $A = (110111)_2$ and $B = (11010)_2$
   (b) when $A = (40731)_8$ and $B = (3134)_8$
   (c) when $A = (110.101)_2$ and $B = (10.01011)_2$

9. Evaluate $A \times B$ in full and $A - B$ quoting the quotient correct to four significant figures:
   (a) when $A = (110.1010)_2$ and $B = (10.01)_2$
   (b) when $A = (75.63)_8$ and $B = (12.24)_8$
   (c) when $A = (0.01001001001 \ldots)_2$ and $B = (11.01)_2$

10. Evaluate, correct to 3B,

$$\sqrt{(1010{\cdot}00111)}$$

11. Using the sign and modulus notation with a binary scale factor of 2^{-5}, as in the text, evaluate the following (a) using 1's complement techniques, and (b) using 2's complement techniques:
    (i) $1{\cdot}10101 + 0{\cdot}01110$
    (ii) $1{\cdot}01101 + 1{\cdot}11001$
    (iii) $0{\cdot}10111 + 1{\cdot}10111$
    (iv) $0{\cdot}11011 + 0{\cdot}10010$
    (v) $0{\cdot}01011 - 0{\cdot}10010$
    (vi) $1{\cdot}01111 - 1{\cdot}00101$

    In all cases indicate when the results overflow.

Chapter 2
The Algebra of Classes

Introduction

There are many different methods by which the design of logic circuits may be approached. The basic mathematical techniques form part of a subject known as Boolean algebra. This is an abstract mathematical system which, starting with a given set of axioms or postulates, allows various theorems to be proved. In this sense it is very similar to elementary geometry.

However, geometry has the advantage that the theorems are about well-known objects such as points, lines, and circles. Thus, to allow an intuitive approach to Boolean algebra, the algebra of classes or sets will be developed first. The ideas considered in this chapter will then form a background to the formal theory and it will be seen that the algebra of classes does in fact form a Boolean algebra.

The subject matter will already be very familiar to many readers whose mathematical training has been based on 'modern' mathematics. To these people the examples will seem particularly easy. However, it has been written assuming no previous knowledge and it is hoped will provide the necessary background for the other readers whose mathematics has been of the 'traditional' kind.

Classes or Sets

The two terms 'class' or 'set' are used here with identical meanings and refer to any well-defined collection of objects. These terms will not be given precise definitions and are best illustrated by examples.

(a) The positive integers.

(b) The letters of the alphabet.

(c) The readers of this book.

(d) The real solutions of the equation $x^2 + 1 = 0$.

(e) The three letters A, B, C.

(f) The numbers 1, 3, 5, 7, ...

(g) The set consisting of the single person George Boole.

The sets in the above examples are defined in two different ways. In examples (a) to (d) the properties of the objects or the rules to form the set are given. In (e) to (g), however, the members of each set are stated explicitly. Either method is satisfactory, the important thing being that it must always be possible to decide whether or not any given object belongs to the set under consideration. Some sets may be defined in either way, for example, the numbers 1, 3, 5, 7, . . . of (f) could be described as the set of positive odd integers.

It is convenient to have a standard notation for defining sets. This may take the form of

$$x = \{1, 3, 5, 7, \ldots\}$$

when all the elements are listed or sufficient of them to enable all the missing ones to be deduced, or the form

$$x = \{X \mid X \text{ is a positive odd integer}\}$$

when the property is stated. This latter is read as 'x is the set of objects X, such that X is a positive odd integer'. Thus, if any particular number X is a positive odd integer, then X is a member of the set x. This *relation* between an element and a set is written $X \in x$. The symbol $\in$ means 'is a member of'. From the idea of a set, it must always be possible to determine whether or not, for any given element X, $X \in x$ is a valid relation. If X is not a member of the set x, the symbol $\notin$ is used. So that for the set x defined above

$$3 \in x$$

but

$$1000 \notin x$$

Two of the sets defined in the examples deserve further attention. Firstly, example (d), which may be defined as

$$x = \{X \mid X^2 + 1 = 0 \text{ and } X \text{ is real}\}$$

has no members since neither solution of $X^2 + 1 = 0$ is real. Such a set with no members is known as the *null set* and is usually denoted by the letter O. Other notations are the figure 0 and the symbol $\emptyset$.

In example (g), the set contains precisely one member and is known as a *unit set*. The set consisting of the single member X is of course $\{X\}$. Note carefully that X and $\{X\}$ are mathematically entirely different. X is an *object* while $\{X\}$ is a *set* that happens to consist only of the single member X. It is possible, however, for a set to have other sets as members. Thus, if

$$a = \{A, B, C\}$$
$$b = \{D, E\}$$

and

$$c = \{F\}$$

it is quite possible to define a set x as

$$x = \{a, b, c\}$$

that is to say

$$x = \{\{A, B, C\}, \ \{D, E\}, \{F\}\}$$

The algebra of classes that is developed in this chapter operates only on the classes or on the sets themselves and not on the individual members that make up the set.

Two sets x and y are said to be *equal*, i.e. $x = y$, if and only if x and y contain exactly the same elements. As an illustration, consider the two sets

$$x = \{X \mid X^2 = 1\}$$
$$y = \{-1, 1\}$$

These both contain the same two members -1 and 1 and therefore $x = y$. The symbol $\neq$ may be used for two sets which do not have the same elements.

If a set z consists of elements all of which are members of another set y, then z is a *subset* of y which is written as $z \subset y$. For example, the set $z = \{1\}$ is a subset of y defined above. The relation between a subset and the set is also known as *inclusion*, so that z is included in y. The difference between the relations 'is a member of' ($\in$) and 'is included in' ($\subset$) should be carefully noted. The first is between an element and a set, and the second between two sets. Thus $1 \in y$ and $z \subset y$ are correct statements, but $\{1\}$ is not a member of y and 1 is not included in y since $\{1\}$ is a set not an element while 1 is a number and cannot therefore be a subset of y.

The set of all objects under discussion in any particular application or example is known as the *universal set* and will be denoted by the letter I. Other symbols in common use are U, i, and the figure 1. The universal set should be defined at the start of any example, since the choice of universal set may affect the result. Thus in considering the set

$$x = \{X \mid X^2 + 1 = 0\}$$

if I is the set of real numbers then x is the null set, while if I is the set of complex numbers $x = \{-i, i\}$. Once the universal set for the particular discussion is decided, any other set being used must be a subset of it.

Venn Diagrams

It is sometimes very convenient to have a pictorial representation of sets and the relationships that exist between the various sets or subsets. Such an illustration is provided by a Venn diagram, in which any set is represented by the interior of a closed plane curve. Let

$$x = \{1, 2, 3, 4\}$$
$$y = \{3, 4, 5, 6\}$$
$$z = \{3, 4\}$$

then these three sets may be shown in the Venn diagram of Figure 2.1.

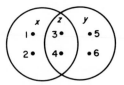

Figure 2.1

Generally, individual elements are not shown in the diagram, and the set is simply represented by the area inside the circle or other simple closed curve. The magnitude of the area does not give any indication of the number of elements contained in the set and it may of course happen that some particular area represents the null set with no members at all. Another possibility is that a set may be represented by the interior of two or more curves. Thus the set

$$w = \{1, 2, 3, 4, 5, 6\}$$

would consist of the interior of both circles in Figure 2.1.

In a Venn diagram, a subset z of a set y is shown by an area lying within another area. So that, in Figure 2.1,

$$z \subset x$$

and also

$$z \subset y$$

The universal set may be considered as the interior of some outer boundary curve, often drawn as a rectangle, and any other set under consideration, which must necessarily be a subset of I, is then an area within this region. Care should be taken in drawing the Venn diagram, so that the most general figure is used and that the correct relationships hold between the various subsets. In Figure 2.2, let the universal set I be the set of all integers, x the set of positive even integers, and y the set of negative integers. Then it is clear that x and y have no common subset and it is best to draw the areas x and y in the diagram so that they do not overlap.

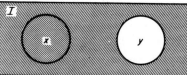

Figure 2.2

It will be possible to use Venn diagrams to demonstrate various theorems relating to logic circuits, but it should be remembered that such diagrams do not constitute a rigorous proof but act only as an illustration and an intuitive verification. They are very convenient for representing the properties of up to three sets or classes. With a little care, a Venn diagram to represent four general related sets can be drawn. Unfortunately, however, it is not possible to draw a completely general relationship between five or more sets on a flat piece of paper and therefore Venn diagrams lose their pictorial advantage for large numbers of sets.

Set Operations

As in ordinary algebra, variables may be combined by operations such as addition and multiplication, so in the algebra of classes there are operations that act on sets. The simplest of these is known as *complementation* and is performed on a single set. Such an operation with one operand is termed *singulary*. The complement of a set y is denoted by $\bar{y}$ and is defined as the set of all the elements of the universal set I that are not members of y. In Figure 2.2, this is represented by the complete shaded area, that is all integers that are not negative. Therefore in this instance the set $\bar{y}$ consists of all the positive integers together with zero.

The other set operations of importance are *binary* and act on two operands. The *union* of two sets x, y is written $x \cup y$ and is the set of all the elements that are either in set x or in set y or in both sets. For the example of Figure 2.1, this consists of the set $w = \{1, 2, 3, 4, 5, 6\}$. The *intersection* of two sets x and y, $x \cap y$, is the set of all the elements common to both sets, i.e. members of x and members of y. This is the set z in Figure 2.1. If two sets have no elements in common they are said to be *disjoint*, e.g.

$$x = \{1, 2\}, \quad y = \{3, 4\}$$

and then

$$x \cap y = O, \text{ the null set}$$

The names given to the symbols for these two set operations are cap ($\cap$) and cup ($\cup$), and a convenient mnemonic to remember is that $\cup$ stands for 'union'. It may be found easier, however, if the operations are referred to as 'x or y' for $x \cup y$ and 'x and y' for $x \cap y$. These immediately infer the composition of the combined set and they are also the names commonly used in logic design for the similar operations that occur there.

There are several other set operations that can be defined and that would normally be considered in a complete study of the algebra of classes, but these three are sufficient for the purposes of this book. As a summary, they are illustrated as Venn diagrams in Figure 2.3.

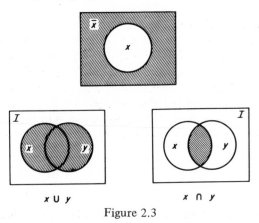

Figure 2.3

Basic Laws

As in algebra there are a number of fundamental laws, so there are similar laws in the algebra of sets. Simple examples from algebra are

$$x + y = y + x$$

and

$$x \cdot y = y \cdot x$$

both of which are examples of the *commutative law*. It should be emphasized that the laws of the algebra of classes are similar to but not identical with the algebraic laws. If any new set operation is encountered, all its laws should be examined carefully. The commutative law holds for the operations of union and intersection which have been defined but other set operations exist for which it is untrue. Of course, it is also not true for the operation of subtraction in algebra, e.g. $x - y \neq y - x$, and multiplication of matrices in matrix algebra, e.g. $A \cdot B \neq B \cdot A$. Each of the basic laws of set algebra, that are given below, can easily be verified by drawing the appropriate Venn diagrams. The reader should convince himself of the truth of each law, and become thoroughly familiar with the permissible operations. The verification of the distributive law (3a) is shown in Figure 2.4.

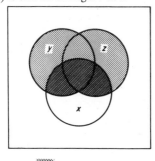
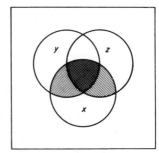

$x \cap (y \cup z)$ Figure 2.4 $(x \cap y) \cup (x \cap z)$

Commutative laws:	(1a) $x \cap y = y \cap x$
	(1b) $x \cup y = y \cup x$
Identity laws:	(2a) $x \cap I = x$
	(2b) $x \cup O = x$
Distributive laws:	(3a) $x \cap (y \cup z) = (x \cap y) \cup (x \cap z)$
	(3b) $x \cup (y \cap z) = (x \cup y) \cap (x \cup z)$
Complement:	(4a) $x \cap \bar{x} = O$
	(4b) $x \cup \bar{x} = I$
Associative laws:	(5a) $x \cap (y \cap z) = (x \cap y) \cap z = x \cap y \cap z$
	(5b) $x \cup (y \cup z) = (x \cup y) \cup z = x \cup y \cup z$
De Morgan's laws:	(6a) $\overline{x \cap y} = \bar{x} \cup \bar{y}$
	(6b) $\overline{x \cup y} = \bar{x} \cap \bar{y}$
Idempotent laws:	(7a) $x \cap x = x$
	(7b) $x \cup x = x$
Operations with O and I:	(8a) $x \cap O = O$
	(8b) $x \cup I = I$

The above laws each occur as pairs. If, in any law, the operations $\cap$ and $\cup$ are interchanged and I and O also interchanged, then the result is also a law of the algebra of classes. This is a consequence of the *principle of duality* which will be discussed more fully in a later chapter.

The relation of inclusion may be related to the operations of union and intersection by noting that the three relations

(a) $x \subset y$
(b) $x \cap y = x$
(c) $x \cup y = y$

each express the identical result that x is a subset of y as shown in Figure 2.5.

These are three fundamental properties of inclusion which may be stated in the following laws.

(1) The reflexive law: $x \subset x$.

(2) The antisymmetric law: if $x \subset y$ and $y \subset x$, then $x = y$.

(3) The transitive law: if $x \subset y$ and $y \subset z$, then $x \subset z$.

It is an example of an ordering relation, but it is only a partial ordering since it does not order every pair of sets. That is to say, in many cases, neither of the

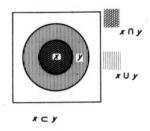

$$x \subset y$$

Figure 2.5

relations $x \subset y$ or $y \subset x$ holds. In this respect it differs from the well-known ordering relation of algebra $\leqslant$ which is a total order relation. The three laws given above hold for inequalities but for any two numbers x, y it is always true that either $x \leqslant y$ or $y \leqslant x$.

Numerical Consistency

There are some applications of the algebra of classes that require a knowledge of the actual number of members of a given set. The number of elements in a set x will be denoted by $F(x)$ and called the *class frequency* of x. If a set is to represent a collection of objects in a real physical situation, then it is clear that the class frequency must not be negative. If a collection of numerical data is given with information about several sets, then the class frequencies of all sets and subsets must be greater or equal to zero. If the data imply that any class frequency is negative, then the data must be inconsistent.

For three variables the smallest areas in a Venn diagram correspond to terms of the form $x \cap y \cap z$ and there are $8 (= 2^3)$ such areas. In general, for n variables, there are 2^n different minimal areas consisting of the intersection of each of the variables or its negation. An independent class frequency can be attached to each of these regions. Thus, for a collection of statistical data, each of 2^n fundamental class frequencies and each area that represents them must be positive if the data are to be consistent.

The information available often does not refer to these minimal areas directly but rather to more complicated regions. It is therefore necessary to calculate the fundamental class frequencies. If the class frequencies $F(x)$ and $F(y)$ are given, what can be deduced regarding $F(x \cap y)$? The answer is that the minimum and maximum possible values of $F(x \cap y)$ can be obtained. The minimum value occurs if x and y are disjoint so that $x \cap y$ is null, and the maximum will occur if either x or y is a subset of the other. Hence

$$0 \leqslant F(x \cap y) \leqslant \text{minimum of } F(x) \text{ and } F(y)$$

If $F(x \cup y)$ is also known, then an exact result may be obtained. Figure 2.6 shows the Venn diagram for two variables. From this it can be seen that, if

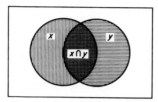

Figure 2.6

the area of x is added to the area of y, then the common region $x \cap y$ is included twice, thus

$$\text{area } x + \text{area } y = \text{area } (x \cup y) + \text{area } (x \cap y)$$

Hence

$$F(x \cup y) = F(x) + F(y) - F(x \cap y)$$

This is a general result and holds for the class frequencies of any two arbitrary sets. A particular case of importance is when the sets x and y are disjoint, i.e. have no elements in common. Then

$$x \cap y = 0 \text{ and hence } F(x \cap y) = 0$$

giving

$$F(x \cup y) = F(x) + F(y)$$

This is the well-known result of probability theory, i.e. the total probability of two mutually exclusive events is the sum of the probabilities of each event.

The general result may be extended to three or more sets. For any three sets $x, y, z,$

$$F(x \cup y \cup z) = F(x) + F(y) + F(z) - F(x \cap y) - F(y \cap z)$$
$$- F(z \cap x) + F(x \cap y \cap z)$$

In many numerical cases, the values of the various class frequencies and determination of consistency can best be obtained by filling in the details on a Venn diagram, as in the following example.

Example. A library produced the following statistical information regarding its book collection. 80% of all the books are written in English, 40% are on technical subjects, and 60% out on loan. A further breakdown of information showed that 35% of the books are on technical subjects in English, 27% are technical books on loan, 53% are books in English on loan, and 25% are technical books in English on loan. Show that the information is consistent and determine (i) the percentage of technical books not on loan; (ii) the percentage of the foreign books which are on technical subjects.

In this example, the universal set I is the total number of books in the library which expressed as a percentage must be 100%.

Let

x be the set of technical books
y be the set of books written in English
z be the set of books out on loan

The information given on class frequencies is then

		%
$F(I)$,	total books	100
$F(x)$,	technical books	40
$F(y)$,	books in English	80
$F(z)$,	books on loan	60
$F(x \cap y)$,	technical books in English	35
$F(x \cap z)$,	technical books on loan	27
$F(y \cap z)$,	books in English on loan	53
$F(x \cap y \cap z)$,	technical books in English on loan	25

Draw first a Venn diagram for three sets x, y, z. The only information that is given for a minimal area is $F(x \cap y \cap z) = 25$ and this is entered on the diagram. This immediately allows the value of $F(x \cap y \cap \bar{z})$ to be calculated, since from the diagram

$$F(x \cap y \cap \bar{z}) = F(x \cap y) - F(x \cap y \cap z)$$
$$= 35 - 25 = 10$$

Similarly

$$F(x \cap \bar{y} \cap z) = 27 - 25 = 2$$
$$F(\bar{x} \cap y \cap z) = 53 - 25 = 28$$

and these are entered on the diagram as in Figure 2.7. From the diagram, it is easy to see how to continue. Three of the four regions which constitute the set x are now determined and $F(x)$ is also known. Therefore

$$F(x \cap \bar{y} \cap \bar{z}) = 40 - (10 + 25 + 2)$$
$$= 40 - 37 = 3$$

and similarly for the other regions, to give the completed Figure 2.8.

All the regions are positive so that the data are consistent. The two pieces of information required are firstly $F(x \cap \bar{z})$ which may be read from the Venn diagram as 3 + 10 = 13. Thus 13% of the technical books are not on loan.

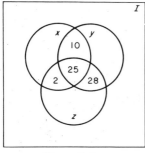

Figure 2.7

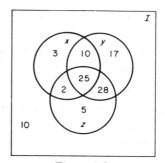

Figure 2.8

The second part requires two steps. The number of foreign books (i.e. not English) on technical subjects is $F(x \cap \bar{y}) = 2 + 3 = 5$. But this is the percentage of all books, not the percentage of the foreign books. Now $F(\bar{y})$, the percentage of foreign books, equals $100 - 80 = 20$. Hence the percentage of the foreign books on technical subjects is $(5/20) \times 100 = 25$.

Suggestions for Further Reading

Birkhoff, G., and Maclane, S., 1953, *A Survey of Modern Algebra,* 2nd edn., Macmillan, New York.

Halmos, P. R., 1960, *Naïve Set Theory,* Van Nostrand, London.

Lipschutz, S., 1964, *Set Theory and Related Topics,* Schaum, New York.

Mansfield, D. E., and Bruckheimer, M., 1965, *Background to Set and Group Theory,* Chatto and Windus, London.

Stoll, R. R., 1961, *Sets, Logic and Axiomatic Theories,* Freeman, San Francisco and London.

—— 1963, *Introduction to Set Theory and Logic,* Freeman, San Francisco and London.

Exercises

1. Write out in full the members of the following sets:
   (i) $\{X \mid X^2 = 9\}$
   (ii) $\{Y \mid Y$ is not a consonant$\}$
   (iii) $\{Z \mid Z > 0$ and $Z \leqslant 0\}$
   (iv) $\{X \mid X$ is the capital of France$\}$
   (v) $\{Y \mid Y + X = 3$ and $2Y - X = 3\}$

2. Let the universal set be the set of all integers
   p the set of positive integers
   e the set of even numbers
   o the set of odd numbers
   z the single number zero
   Determine which of the following statements are correct and give examples
   to disprove false statements:

   (a) $o \subset p$ (b) $p \subset e \cup o$
   (c) $p \cap e = p$ (d) $e \cap o = I$
   (e) $e \cap o = O$ (f) $\{X \mid X^2 = 9\} \subset p$
   (g) $\{2\} \in e$ (h) $\bar{e} = o$
   (i) $(e \cup o) \cap p = e$ (j) $z \in \bar{p}$
   (k) $e \cap o \cap \{z\} = I$ (l) $\bar{e} \cap \bar{o} \cap \bar{z} = I$

3. State which of the following are correct, giving reasons for your answers:
   (a) $\{1, 2, 3\} = \{3, 2, 1\}$
   (b) $\{1, 2, 3, 4\} \subset \{1, 2, 3, 5\}$
   (c) $\{1, 2, 1, 2, 3\} = \{1, 2, 3\}$
   (d) $\{1, 2\} \in \{1, 2, \{1, 2, 3\}, \{\{1, 2\}\}\}$

   Show that
   $$\{\{x\}, \{x, y\}, \{x, y, z\}\} = \{\{a\}, \{a, b\}, \{a, b, c\}\}$$
   if and only if $x = a, \quad y = b, \quad z = c.$

4. Given two sets x and y which are not disjoint, indicate the following sets
   on a Venn diagram:
   (a) $x \cap (y \cup \bar{y})$
   (b) $x \cap (y \cap \bar{y})$
   (c) $x \cup y \cup \bar{x} \cup \bar{y}$
   (d) $\bar{x} \cap \bar{y}$
   (e) $\overline{x \cup y}$

5. Draw the correct Venn diagram for three sets x, y, z to satisfy the following relations:
   (a) $x \subset y$, $y \subset z$
   (b) $x \subset z$, $y \cap z = 0$
   (c) $x \subset y$, $x \cap z \neq 0$, $z \not\subset y$
   (d) $x \cap y = 0$, $y \cap z = 0$, $z \cap x = 0$
   (e) $x \cap y \neq 0$, $z \subset x \cap y$

6. The *difference* of two sets x, y is the set of elements which belong to x but do not belong to y and is written $x - y$. Indicate this operation on a Venn diagram and express $x - y$ in terms of complementation and intersection. Determine whether the difference operation is
   (a) commutative
   (b) distributive over intersection
   (c) associative
   Prove the results
   (i) $(x - y) \cap y = 0$
   (ii) $(x - y) \subset x$
   (iii) $x - y$, $x \cap y$, $y - x$ are mutually disjoint
   (iv) $(x - y) \subset \bar{y}$

7. The *symmetric difference* of two sets x, y is the set of all members of x which are not members of y and all members of y which are not members of x. This is written as $x \triangle y$ and $x \triangle y = (x \cap \bar{y}) \cup (\bar{x} \cap y)$. Indicate this operation on a Venn diagram and show that

$$x \triangle y = (x \cup y) - (x \cap y)$$
$$= (x - y) \cup (y - x)$$

   Demonstrate the following results both by the algebra of classes and by drawing Venn diagrams:
   (a) $x \triangle y = y \triangle x$
   (b) $x \triangle (y \triangle z) = (x \triangle y) \triangle z$
   (c) $x \cup (y \triangle z) \neq (x \cup y) \triangle (x \cup z)$
   (d) $x \cap (y \triangle z) = (x \cap y) \triangle (x \cap z)$
   (e) $x \triangle 0 = x$
   (f) $x \triangle x = 0$
   (g) if $x \triangle y = x \triangle z$, then $y = z$

8. The following sets are defined:

$$I = \{1, 2, 3, 4, 5, 6, 7, 8\}$$
$$w = \{1, 2, 3, 4\}$$
$$x = \{1, 3, 5, 7\}$$
$$y = \{2, 4, 6, 8\}$$
$$z = \{3, 4, 5, 6\}$$

Find the members of the following sets:
(a) $w - x$ (b) $w \triangle x$
(c) $w \triangle x \triangle y \triangle z$ (d) $(w - x) - y$
(e) $w - (x \cap y)$ (f) $(x - y) \triangle x \triangle w$
(g) $w \triangle \bar{w}$ (h) $(x \cap y) \triangle (w \cap z)$

9. Prove the following results for the operation of inclusion:
(a) if $x \subset y$, then $x \cap \bar{y} = 0$
(b) if $x \subset y$, then $\bar{x} \cup y = I$
(c) if $x \subset y$, then $x \cup (y - x) = y$
(d) if $x \subset y$, then $\bar{y} \subset \bar{x}$
(e) if $x \cup y = I$, then $\bar{x} \subset y$
(f) if $x \cap y = 0$, then $x \subset \bar{y}$

10. Express the following conditions in terms of set inclusion:
(a) $\bar{w} \cup \bar{x} \cup \bar{y} \cup z = I$
(b) $x \cup y \cup z = x \cup y$
(c) $x \cap y \cap z = x \cap z$
(d) $x \cap (\bar{y} \cup \bar{z}) = 0$
(e) $x \cap \bar{y} = 0$ and $\bar{y} \cup z = I$

11. Prove that, for any three sets x, y, z,

$$F(x \cup y \cup z) = F(x) + F(y) + F(z)$$
$$- F(x \cap y) - F(y \cap z) - F(z \cap x) + F(x \cap y \cap z)$$

Show further that, if $x_1, x_2, \ldots, x_n$ are n sets which are mutually disjoint and satisfy

$$x_1 \cup x_2 \cup \ldots \cup x_n = I$$

then for any set y

$$F(y) = F(x_1 \cap y) + F(x_2 \cap y) + \ldots + F(x_n \cap y)$$

12. The number of members of three sets x, y, z satisfy the following relations:

$$F(x) = 10$$
$$F(y) = 8$$
$$F(z) = 9$$
$$F(x \cap y) = 6$$
$$F(y \cap z) = 4$$
$$F(x \cap z) = 5$$
$$F(x \cap y \cap z) = 3$$
$$F(\bar{x} \cap \bar{y} \cap \bar{z}) = 3$$

Find the total number of objects in the universal set I and show that

$$F(\bar{z}) = F(z)$$

and

$$F(x \cap \bar{z}) = F(x \cap z)$$

13. The following statistics were obtained regarding the population of a certain town:

   55% are male
   30% are children
   60% are employed (for at least 3 hours per week)
   60% of the children are girls
   80% of males are employed
   70% of the children are not employed
   5% of the population are unemployed male adults

Determine the following from these data:

(a) the percentage of the population who are unemployed female adults
(b) the percentage of the female children who are employed
(c) the percentage of females who are adult

14. Draw a general Venn diagram for four non-disjoint sets w, x, y, z. Hence show that the following set of data is inconsistent.

In a school sixth-form, the pupils may study up to three subjects chosen from Mathematics, English, History, and Physics. They must take at least one subject and may not take all four. It was found that from a class of 40 pupils

   29 study Mathematics
   22 study English
   15 study History
   10 study Physics
   10 study English and History

   2 study English and Physics
   1 studies History and Physics
   11 studying Mathematics do not study English or History
   15 studying Mathematics do not study English
   2 study only English
   Everybody who studies Physics also studies Mathematics
Determine the number of pupils in the class that would give a consistent
result and also the number of pupils who study
   (a) one subject only
   (b) exactly two subjects
   (c) exactly three subjects

15. A breakdown of 50 employees of a firm gave the following figures:
      number of graduates   = 10
      number of men   = 23
      number of married persons   = 21
      number of single men   = 7
      number of non-graduate men   = 15
Determine the number of married men employed and the minimum and
maximum possible numbers of married graduates and of single women
non-graduates.

Chapter 3
Boolean Algebra

Boolean Algebra as an Axiomatic System

In the previous chapter, an intuitive approach was adopted to enable the
fundamental ideas to be understood easily. Now a more formal mathematical
system will be presented. This starts with a set of basic axioms or postulates
from which all other theorems of the algebra can be proved by rigorous
methods. This abstract mathematical system is known as Boolean algebra and
is named after the English mathematician George Boole (1815–64) who
developed a symbolic method of studying the relationships between sets and
for investigating logical propositions (see Boole, 1951). The extension to the
algebra of propositions is contained in Chapter 4, while in this chapter it will
be shown that the algebra of classes that was developed earlier satisfies all the
axioms of Boolean algebra. When interpreting the theorems of Boolean algebra
it may prove helpful to use examples from this set algebra and, in particular,
Venn diagrams may be used to illustrate a theorem but will not of course
constitute a formal proof.

There are several different possible axiomatic systems for Boolean algebra.
In choosing a set of axioms for any abstract system, there are a number of considerations to be taken into account. The most important of these is that the
set of axioms should be consistent. A *consistent set* is one in which it is
impossible to prove that some proposition or theorem is true and also to prove
that the same proposition is false. This requirement is obviously necessary if

any useful interpretation of the results of the theory is to apply to practical systems. It is desirable to have simple axioms that also lend themselves to easy proofs of the theorems of the system. It is in fact possible to derive a system of Boolean algebra that uses just *one* axiom. This is very elegant from a mathematical point of view, but from the practical side, however, proofs of theorems become extremely tedious and the whole system is difficult to understand. A further consideration is the independence of the axioms. A set of axioms is *independent* if none of them can be derived from the remaining axioms. If the set of axioms chosen is not independent, it means that one or more of the axioms can be discarded and that the remaining axioms will still suffice to prove all the theorems.

The set of four axioms chosen here originates from Huntington (1904) and is particularly easy to associate with the algebra of classes and with the logic design of switching circuits. It is consistent and the axioms are independent.

A Boolean algebra is a set of elements B with two binary operations . and + for which the following axioms hold.

Axiom A1. The operations . and + are commutative.

Axiom A2. There exist identity elements 0 and 1 in B with respect to each of the operations . and +.

Axiom A3. Each operation is distributive over the other.

Axiom A4. For every element x in B, there exists an inverse element $\bar{x}$ in B such that $x \cdot \bar{x} = 0$ and $x + \bar{x} = 1$.

The operation . is called Boolean product and + is called Boolean sum. The basic axioms give the following equations, when written out for any two elements x, y of B:

(1a) $x \cdot y = y \cdot x$ (1b) $x + y = y + x$

(2a) $x \cdot 1 = x$ (2b) $x + 0 = x$

(3a) $x \cdot (y + z) = (x \cdot y) + (x \cdot z)$ (3b) $x + (y \cdot z) = (x + y) \cdot (x + z)$

(4a) $x \cdot \bar{x} = 0$ (4b) $x + \bar{x} = 1$

The operation . is assumed to take precedence over + in unbracketed expressions.

Care must be taken to avoid confusion between Boolean sum and product and the normal mathematical operations of plus and multiply. Although some of the properties are very similar, others are completely different. Thus the distributive law of (3a) is identical in form with

$$x(y + z) = xy + xz$$

but the dual axiom of (3b) is completely untrue in ordinary algebra. To manipulate Boolean expressions, it is necessary to be quite familiar with the distributive law in both its forms.

If Equations (1)–(4) above are compared with the corresponding numbered equations in Chapter 2 for the laws of the algebra of classes

with union $\cup$ written for +
intersection $\cap$ for .
universal set I for 1
null set O for 0

it will be seen that the algebra of classes satisfies each of the axioms. Hence the algebra of classes is a Boolean algebra under the operations of union and intersection. The remaining laws of Chapter 2 can all be proved for any Boolean algebra from Axioms A1–A4, so that any algebra of classes can be interpreted as a Boolean algebra.

The converse of this is also true. Any Boolean algebra may be interpreted as an algebra of classes for an appropriate universal set.

It can in fact be shown that every Boolean algebra is isomorphic to an algebra of classes. An *isomorphism* is a very close relationship between two systems that preserves 'sums and products'. Thus, if B is a Boolean algebra with a sum operation + and a product operation ., then there is an algebra of classes B' with a sum operation $\cup$ and a product operation $\cap$. There is a one to one correspondence between any element x of B and an element x' of B' such that for any elements x and y of B

$$(x + y)' = x' \cup y'$$
$$(x . y)' = x' \cap y'$$

Also the zero and unity elements of B correspond to the zero and unity elements of B'

$$0' = O$$
$$1' = I$$

The result of this isomorphism is that any theorem discovered from the Venn diagrams or the algebra of classes remains true in the corresponding Boolean algebra and can be proved from the set of axioms.

Theorems of Boolean Algebra

Proofs will now be given of some theorems in Boolean algebra to illustrate the method.

Firstly the principle of duality must hold. That is, if in any valid algebraic identity or expression the operations of . and + and the elements 0 and 1 are interchanged, then the resultant expression is also valid. This is self-evident as the axioms are symmetrical with respect to the two operations and two identity elements.

When the principle of duality is applied to an algebraic expression, the result is known as the *dual* of the former. Applying the principle again, it is clear that the dual of the latter gives the original expression. Every theorem can therefore be stated in two forms which are dual statements. The various steps in proofs of a pair of dual theorems will also be dual statements. There is thus no need to give proofs for both theorems.

Theorem 3.1. The inverse element $\bar{x}$ associated with x is unique.

Proof. Suppose there exist two inverse elements a and b corresponding to an element x so that

$$x + a = 1, \quad x \cdot a = 0$$
$$x + b = 1, \quad x \cdot b = 0$$

Then

$$
\begin{aligned}
a &= a \cdot 1 & \text{(A2)} \\
&= a \cdot (x + b) & \\
&= (a \cdot x) + (a \cdot b) & \text{(A3)} \\
&= (x \cdot a) + (a \cdot b) & \text{(A1)} \\
&= 0 + (a \cdot b) & \\
&= (x \cdot b) + (a \cdot b) & \\
&= (b \cdot x) + (b \cdot a) & \text{(A1)} \\
&= b \cdot (x + a) & \text{(A3)} \\
&= b \cdot 1 & \\
&= b & \text{(A2)}
\end{aligned}
$$

Thus, if a and b are each inverses of x, then $a = b$. Note that, for an inverse, it is necessary for both of the equations

$$x + a = 1 \qquad \text{and} \qquad x \cdot a = 0$$

to be satisfied for a to be the inverse of x. For example,

$$
\begin{aligned}
x + (\bar{x} + y) &= (x + \bar{x}) + y \\
&= 1 + y \\
&= 1
\end{aligned}
$$

thus satisfying the first equation. But

$$
\begin{aligned}
x \cdot (\bar{x} + y) &= x \cdot \bar{x} + x \cdot y \\
&= 0 + x \cdot y \\
&= x \cdot y \\
&\neq 0
\end{aligned}
$$

Theorem 3.2. Both of the operations . and + are associative:

(5a) $x \cdot (y \cdot z) = (x \cdot y) \cdot z$

(5b) $x + (y + z) = (x + y) + z$

This theorem on associativity states that the same result is obtained whichever order the operations are performed. Thus the brackets are unnecessary and the expressions may be written as $x \cdot y \cdot z$ and $x + y + z$ respectively.

The proof of this theorem, although it uses a similar method to the previous theorem, is quite complicated and will therefore be omitted. It is, however, necessary to insert the theorem at this point in the text, as the result will be required in following proofs. It is an essential requirement of an axiomatic system that theorems should be proved in a logical order, and should not use in their proof any result that has not yet been obtained.

Theorem 3.3. For every element x in B

$$\overline{(\overline{x})} = x$$

Proof.

$$\overline{x} + x = 1 \qquad \text{and} \qquad \overline{x} \cdot x = 0 \qquad \text{(A4)}$$

so that x is the unique inverse of $\overline{x}$ (Theorem 3.1), i.e.

$$\overline{(\overline{x})} = x$$

Theorem 3.4. For any elements x, y in B

$$\overline{x \cdot y} = \overline{x} + \overline{y}; \qquad \overline{x + y} = \overline{x} \cdot \overline{y} \qquad \text{(De Morgan's laws)}$$

Proof. The method of proof is to show that $\overline{x} + \overline{y}$ is the inverse of $x \cdot y$. Thus it is necessary to show that

(i) $(\overline{x} + \overline{y}) + x \cdot y = 1$
(ii) $(\overline{x} + \overline{y}) \cdot (x \cdot y) = 0$

Firstly

$$
\begin{aligned}
(\overline{x} + \overline{y}) + x \cdot y &= (\overline{x} + \overline{y} + x) \cdot (\overline{x} + \overline{y} + y) & \text{(A3)} \\
&= (\overline{x} + x + \overline{y}) \cdot (\overline{x} + \overline{y} + y) & \text{(A1)} \\
&= (1 + \overline{y}) \cdot (\overline{x} + 1) & \text{(A4)} \\
&= 1 \cdot 1 & \text{(A2)} \\
&= 1 & \text{(A2)}
\end{aligned}
$$

Secondly

$$
\begin{aligned}
(\overline{x} + \overline{y}) \cdot (x \cdot y) &= \overline{x} \cdot x \cdot y + \overline{y} \cdot x \cdot y & \text{(A3)} \\
&= (\overline{x} \cdot x) \cdot y + (\overline{y} \cdot y) \cdot x & \text{(5a)} \\
&= 0 \cdot y + 0 \cdot x & \text{(A4)} \\
&= 0 + 0 & \text{(A2)} \\
&= 0 & \text{(A2)}
\end{aligned}
$$

Thus $(\bar{x} + \bar{y})$ is the inverse of $x \cdot y$ and, since by Theorem 3.1 this inverse must be unique, it must be equal to $\overline{x \cdot y}$, i.e.

$$\bar{x} + \bar{y} = \overline{x \cdot y}$$

These two laws have extensions to any number of variables in the form

$$\overline{p \cdot q \cdot \ldots \ldots \cdot y \cdot z} = \bar{p} + \bar{q} + \ldots \ldots + \bar{y} + \bar{z}$$
$$\overline{p + q + \ldots \ldots + y + z} = \bar{p} \cdot \bar{q} \cdot \ldots \ldots \cdot \bar{y} \cdot \bar{z}$$

These two laws of De Morgan are an example of the principle of duality in a more general form. This may be stated as follows.

If E is any Boolean expression involving the operations of $.$, $+$, $^{-}$, and E^* is obtained by interchanging every occurrence of $.$ and $+$ and replacing each Boolean variable by its inverse, then $E = \overline{E^*}$.

This gives a convenient method of finding the inverse of compound Boolean expressions. For example, suppose the inverse of $f = (\bar{x} + y) \cdot (x + y + z \cdot \bar{x}) + \bar{x} \cdot \bar{y}$ is required. Parentheses should be inserted where necessary to maintain the correct grouping of operators. Thus

$$f = [(\bar{x} + y) \cdot (x + y + (z \cdot \bar{x}))] + (\bar{x} \cdot \bar{y})$$

and hence

$$\bar{f} = f^* = [(x \cdot \bar{y}) + (\bar{x} \cdot \bar{y} \cdot (\bar{z} + x))] \cdot (x + y)$$
$$= [x \cdot \bar{y} + \bar{x} \cdot \bar{y} \cdot (\bar{z} + x)] \cdot (x + y)$$

Theorem 3.3, together with De Morgan's laws, gives a convenient method of eliminating either the product operation $.$ or the sum operation $+$ from any Boolean expression.

Example. Express $x \cdot \bar{y} \cdot z + x \cdot y \cdot \bar{z}$ in terms of the $.$ and $^{-}$ operations only and in terms of the $+$ and $^{-}$ only.

$$x \cdot \bar{y} \cdot z + x \cdot y \cdot \bar{z} = \overline{\overline{x \cdot \bar{y} \cdot z + x \cdot y \cdot \bar{z}}} \qquad \text{(Theorem 3.3)}$$
$$(. \text{ only}) \qquad = \overline{\overline{x \cdot \bar{y} \cdot z} \cdot \overline{x \cdot y \cdot \bar{z}}} \qquad \text{(Theorem 3.4)}$$
$$= \overline{(\bar{x} + y + \bar{z}) \cdot (\bar{x} + \bar{y} + z)} \qquad \text{(Theorem 3.4)}$$
$$(+ \text{ only}) \qquad = \overline{\bar{x} + y + \bar{z}} + \overline{\bar{x} + \bar{y} + z} \qquad \text{(Theorem 3.4)}$$

The remaining laws introduced in Chapter 2, which may be proved in a similar manner to the above, are

(7a) $x \cdot x = x$ (7b) $x + x = x$
(8a) $x \cdot 0 = 0$ (8b) $x + 1 = 1$

Other useful relations are

(9a) $x + (x \cdot y) = x$ (9b) $x \cdot (x + y) = x$
(10) $(x \cdot y) + (y \cdot z) + (z \cdot x) = (x + y) \cdot (y + z) \cdot (z + x)$
(11a) $\bar{0} = 1$ (11b) $\bar{1} = 0$

Any of the previous theorems may be used in simplifying Boolean expressions. As will have been noticed in the proofs of the theorems, it is often necessary to expand an expression before it can be simplified.

Example. Simplify the expression $\bar{x} \cdot y + y \cdot \bar{z} + \bar{x} \cdot z$:

$$\bar{x} \cdot y + y \cdot \bar{z} + \bar{x} \cdot z = (\bar{x} \cdot y \cdot 1) + y \cdot \bar{z} + \bar{x} \cdot z$$
$$= \bar{x} \cdot y \cdot z + \bar{x} \cdot y \cdot \bar{z} + y \cdot \bar{z} + \bar{x} \cdot z$$
$$= (\bar{x} \cdot z) \cdot (y + 1) + (y \cdot \bar{z}) \cdot (\bar{x} + 1)$$
$$= \bar{x} \cdot z + y \cdot \bar{z}$$

There are two operations which are not permissible. These are as follows.

(i) Equal terms cannot be removed from each side of Boolean equations. Thus is was proved above that

$$\bar{x} \cdot y + y \cdot \bar{z} + \bar{x} \cdot z = \bar{x} \cdot z + y \cdot \bar{z}$$

If equal terms could be removed, this would result in $\bar{x} \cdot y = 0$ which is clearly incorrect.

(ii) In a similar manner, cancellation of factors is prohibited, i.e. if $A \cdot B = A \cdot C$ it is not correct to deduce that $B = C$, e.g. $x \cdot (\bar{x} + y) = x \cdot y$.

Order Relation

In the algebra of classes there is an inclusion relation $x \subset y$ which indicates that x is a subset of y. Since a Boolean algebra is isomorphic to the algebra of classes, a corresponding type of relation must exist in Boolean algebra. The partial order relation $x \leqslant y$ in a Boolean algebra B is defined to be equivalent to the equation $x \cdot \bar{y} = 0$. From this definition it is easy to show that this order relation obeys the following laws.

(1) Reflexive law. $x \leqslant x$.
(2) Antisymmetric law. If $x \leqslant y$ and $y \leqslant x$, then $x = y$.
(3) Transitive law. If $x \leqslant y$ and $y \leqslant z$, then $x \leqslant z$.

These laws are also satisfied by the inclusion relation of the algebra of classes. The other properties of inclusion which carry over to Boolean algebra are the consistency principle that

$$x \leqslant y, \quad x \cdot y = x, \quad x + y = y$$

each express the same relationship, and that

$$x \leqslant y \quad \text{if and only if} \quad \bar{y} \leqslant \bar{x}$$

There are four possible ways in which two elements x and y in a Boolean algebra B may be related by the order or inclusion relation.

(1) $x \leqslant y$ and $y \leqslant x$ leading to $x = y$ (the antisymmetric law above).

(2) $x \leqslant y$ but not $y \leqslant x$. This may be written as $x < y$ (i.e. strict inclusion $x \leqslant y$ but $x \neq y$).

(3) $y \leqslant x$ but not $x \leqslant y$. This may be written as $x > y$.

(4) Neither $x \leqslant y$ nor $y \leqslant x$ holds. x and y are then called *incomparable*.

It is this existence of incomparable elements that distinguishes the inclusion relation from the inequality relation in conventional algebra.

Canonical or Normal Forms

The terms *Boolean product* and *Boolean sum* have already been applied to the operations . and +. It is also useful to define other expressions such as polynomial, factor, etc., for Boolean algebra.

Constant. Any single symbol which represents a particular element of a Boolean algebra, e.g. 0 and 1. Use letters a, b, c at the beginning of the alphabet to represent constants.

Variable. Any symbol x, y, etc., used to represent an arbitrary or unknown element.

Literal. A variable x or its inverse $\bar{x}$.

Boolean function. This may be defined as follows. Firstly $F(x) = b$, where b is a Boolean constant, is a function. $F(x) = x$ for all x, where x is a variable is a function. Any expression that can be formed from these two functions by applying the operations $^{-}$, +, and . a finite number of times is a Boolean function.

Monomial. A Boolean product of two or more variables x, y or their inverses $\bar{x}$, $\bar{y}$, for example $x . \bar{y} . z$. The single literal x standing alone may be considered as a special case of a monomial.

Polynomial. A Boolean sum of monomials (each individual monomial is called a *term*), for example

$$x . y + \bar{x} . z + x . y . z$$

This is also called a *disjunctive normal form*.

Linear factor. A Boolean sum of simple literals, for example

$$x + \bar{y} + z$$

A polynomial may always be factorized into a set of linear factors to give an expression that consists of a product of sum terms, for example the polynomial

$$\begin{aligned} w . y + w . z + x . y + x . z \\ = w . (y + z) + x . (y + z) \\ = (w + x) . (y + z) \end{aligned}$$

This is called a *conjunctive normal form*.

Any Boolean function can always be written as a disjunctive normal form or a conjunctive normal form. In a normal form, the inverse symbol $^-$ must only occur over single elements and may not act as a bracket; for example $x \cdot y + \overline{x \cdot z}$ is not a normal form. It can be converted to normal form by applying De Morgan's law to the inverse term giving

$$x \cdot y + \bar{x} + \bar{z}$$

The above normal forms do not give a unique representation of a Boolean function. It is very convenient to define a unique expression for any Boolean function (apart from the dual expression). This is done by means of the *special* normal forms.

The special disjunctive normal form is a disjunctive normal form (i.e. a sum of products or polynomial) in which every term contains *all n* variables either with or without inversion, and no two terms are identical.

Every Boolean function which contains no constants is equal to a unique expression in special disjunctive normal form.

To obtain the special disjunctive normal form, the first step is to remove any inverse connectives that are acting as brackets by applying De Morgan's law as in the previous example. Continue this process until the only $^-$ signs appear over single variables. Then reduce the expression to a polynomial by using the distributive law.

If any particular term does not involve all the primitive elements, it must be expanded by forming the product with $x_i + \bar{x}_i$ which is equal to 1 and will therefore leave the value unchanged. For example, in an expression of the three variables x, y, z suppose one term is $x \cdot z$. Then

$$\begin{aligned} x \cdot z &= (x \cdot z) \cdot (y + \bar{y}) \\ &= x \cdot z \cdot y + x \cdot z \cdot \bar{y} \\ &= x \cdot y \cdot z + x \cdot \bar{y} \cdot z \end{aligned}$$

Continue this until every term contains products of all variables. Finally eliminate duplicate terms using Equation (7b) $x + x = x$. The result will then be in special disjunctive normal form.

Example. Express in special disjunctive normal form the expression in three variables $x + \overline{\overline{x + y} + \overline{y + z}} + x \cdot y \cdot \bar{z}$.

$$\begin{aligned} f &= x + \overline{\overline{x + y} + \overline{y + z}} + x \cdot y \cdot \bar{z} \\ &= x + \overline{\overline{x + y}} \cdot \overline{\overline{y + z}} + x \cdot y \cdot \bar{z} \\ &= x + (x + y) \cdot (y + z) + x \cdot y \cdot \bar{z} \\ &= x + (x \cdot y + x \cdot z + y \cdot y + y \cdot z) + x \cdot y \cdot \bar{z} \\ &= x + x \cdot y + x \cdot z + y + y \cdot z + x \cdot y \cdot \bar{z} \\ &= x \cdot (1 + y + z + y \cdot \bar{z}) + y \cdot (1 + z) \\ &= x + y \\ &= x \cdot (y + \bar{y}) \cdot (z + \bar{z}) + (x + \bar{x}) \cdot y \cdot (z + \bar{z}) \end{aligned}$$

$$= x \cdot y \cdot z + x \cdot y \cdot \bar{z} + x \cdot \bar{y} \cdot z + x \cdot \bar{y} \cdot \bar{z} + \bar{x} \cdot y \cdot z + x \cdot y \cdot \bar{z} + \bar{x} \cdot y \cdot z$$
$$+ \bar{x} \cdot y \cdot \bar{z}$$
$$= \bar{x} \cdot y \cdot \bar{z} + \bar{x} \cdot y \cdot z + x \cdot \bar{y} \cdot \bar{z} + x \cdot \bar{y} \cdot z + x \cdot y \cdot \bar{z} + x \cdot y \cdot z$$

There are 2^n possible distinct terms in the special disjunctive normal form of n variables. If they are all present, this may be called the complete disjunctive normal form and the value of the expression is 1. On a Venn diagram each term corresponds to the smallest areas into which the diagram is divided. The sum of all these regions gives the universal set I. Thus the complete disjunctive normal form in three variables is

$$\bar{x} \cdot \bar{y} \cdot \bar{z} + \bar{x} \cdot \bar{y} \cdot z + \bar{x} \cdot y \cdot \bar{z} + \bar{x} \cdot y \cdot z + x \cdot \bar{y} \cdot \bar{z} + x \cdot \bar{y} \cdot z + x \cdot y \cdot \bar{z} + x \cdot y \cdot z$$

(and is equal to 1).

This gives a simple method for writing down immediately the inverse of a function. The inverse of a function must contain precisely those terms which are missing in the special disjunctive normal form. Thus the inverse of the above example must be

$$\bar{x} \cdot \bar{y} \cdot \bar{z} + \bar{x} \cdot \bar{y} \cdot z$$

It can be shown that a Boolean function is completely specified by the values it takes for all combinations of 0 and 1 which may be assigned to the variables. Thus, for example, with two variables

$$F(x, y) = F(0, 0) \cdot \bar{x} \cdot \bar{y} + F(0, 1) \cdot \bar{x} \cdot y + F(1, 0) \cdot x \cdot \bar{y} + F(1, 1) \cdot x \cdot y$$

where the terms $F(0, 0)$, etc., must be either 0 or 1 if the function F does not contain any constants, and hence the special disjunctive normal form is obtained.

It also follows directly from this result that two functions are equal if and only if their special disjunctive normal forms (in the same number of variables) are identical.

The *special conjunctive normal form* is the dual form of the special disjunctive normal form and consists of a conjunctive normal form (i.e. a product of sums) in which every factor contains the sum of all n variables with or without inversion, and no two factors are identical.

The method of procedure in all cases is the dual of the preceding section and again every Boolean function which contains no constants may be written as a special conjunctive normal form. There are 2^n possible distinct factors in this form for n variables. If they are all present, the value of the expression is 0. On a Venn diagram a factor corresponds to a maximum area, being the complete area less one minimum subset. For example $x + y + z$ is the universal set I less the area $(\bar{x} \cdot \bar{y} \cdot \bar{z})$.

Example. Express in special conjunctive normal form the expression in three variables $x + \overline{x + y} + \overline{y + z} + x \cdot y \cdot \overline{z}$.

$$f = x + \overline{\overline{x + y}} + \overline{\overline{y + z}} + x \cdot y \cdot \overline{z}$$
$$= x + \overline{\overline{x + y}} \cdot \overline{\overline{y + z}} + x \cdot y \cdot \overline{z}$$
$$= x + (x + y) \cdot (y + z) + x \cdot y \cdot \overline{z}$$
$$= x \cdot (1 + y \cdot \overline{z}) + (x + y) \cdot (y + z)$$
$$= x + (x + y) \cdot (y + z)$$
$$= (x + x + y) \cdot (x + y + z)$$
$$= (x + y) \cdot (x + y + z)$$
$$= (x + y + z \cdot \overline{z}) \cdot (x + y + z)$$
$$= (x + y + z) \cdot (x + y + \overline{z}) \cdot (x + y + z)$$
$$= (x + y + z) \cdot (x + y + \overline{z})$$

It is very simple to change from one special normal form to another by using De Morgan's law. To convert the above special conjunctive normal form to special disjunctive normal form proceed as follows.

$$(x + y + z) \cdot (x + y + \overline{z}) = \overline{\overline{(x + y + z) \cdot (x + y + \overline{z})}}$$
$$= \overline{\overline{x + y + z} + \overline{x + y + \overline{z}}}$$
$$= \overline{\overline{x} \cdot \overline{y} \cdot \overline{z} + \overline{x} + \overline{y} + z}$$

Taking the inverse, this gives

$$x \cdot y \cdot z + x \cdot y \cdot \overline{z} + x \cdot \overline{y} \cdot z + x \cdot \overline{y} \cdot \overline{z} + \overline{x} \cdot y \cdot z + \overline{x} \cdot y \cdot \overline{z}$$

which is the same result obtained previously.

Boolean Algebra with Two Elements

The number of elements in any Boolean algebra B has so far not been specified. Every Boolean algebra must have more than one element, since it contains elements 0 and 1 from the axioms. A very important Boolean algebra is in fact one that contains only these two elements 0 and 1. The product and sum operations are defined by

$$0 \cdot 0 = 0 \qquad 0 + 0 = 0$$
$$0 \cdot 1 = 1 \cdot 0 = 0 \qquad 0 + 1 = 1 + 0 = 1$$
$$1 \cdot 1 = 1 \qquad 1 + 1 = 1$$

With these definitions, all the axioms for a Boolean algebra are satisfied.

This Boolean algebra is of particular importance since many physical systems may be considered to have precisely two states, e.g. a switch has the two positions 'on' and 'off'. Thus letting 1 in the Boolean algebra represent 'on' and 0 represent 'off', an algebraic model of a switching circuit may be set up. This type of operation will be the main application of Boolean algebra in the future chapters.

References and Further Reading

Boole, G., 1951, *An Investigation of the Laws of Thought*, reprint, Dover Publications, New York.

Halmos, P. R., 1963, *Lectures on Boolean Algebra*, Van Nostrand, London.

Huntington, E. V., 1904, 'The algebra of logic', *Trans. Am. Math. Soc.*, 5, 288–309.

Stone, M. H., 1936, 'The theory of representations for Boolean algebras', *Trans. Am. Math. Soc.*, **40**, 37–111.

Whitesitt, J. E., 1961, *Boolean Algebra and its Applications*, Addison-Wesley, Reading, Mass.

Exercises

1. Using the axioms of Boolean algebra and the associative law, prove the following results:

   (i)   the idempotent laws   (a) $x + x = x$ (b) $x \cdot x = x$
   (ii) the absorption laws   (a) $x + (x \cdot y) = x$ (b) $x \cdot (x + y) = x$
   (iii)   (a) $x + 1 = 1$ (b) $x \cdot 0 = 0$
   (iv)   (a) $x + \bar{x} \cdot y = x + y$ (b) $x \cdot (\bar{x} + y) = x \cdot y$

2. Prove that, if $x + y = x + z$ and $\bar{x} + y = \bar{x} + z$, then $y = z$.

3. Write out the dual proof of Theorem 3.4 to prove the De Morgan law $\overline{x + y} = \bar{x} \cdot \bar{y}$.

4. Use the general principle of duality to find the inverses of the following functions:

   (a) $x + y \cdot z$

   (b) $x \cdot y + y \cdot z + z \cdot x$

   (c) $\overline{x + y} \cdot \overline{y + z} \cdot \overline{z + x}$

   (d) $\overline{\overline{x + \bar{y} + \bar{z}} + \overline{\bar{x} \cdot \bar{y} \cdot \bar{z}}}$

   The final answer should only contain inverses of individual elements and not of compound expressions.

5. Simplify the following expressions:
   (a) $x . \bar{z} + x . y . z + x . \bar{y} . z$
   (b) $(x + y) . (x + z) . (x + \bar{y})$
   (c) $\overline{x . (\bar{y} + z) + \overline{x . y}}$

6. Prove the transitive law for the partial order relation of Boolean algebra
   that

$$\text{if } x \leqslant y \text{ and } y \leqslant z \text{ then } x \leqslant z$$

7. Prove that the dual of $x \leqslant y$ is $y \leqslant x$.

8. Prove by induction that a Boolean function of n variables $F(x_1, x_2, \ldots, x_n)$
   can be expressed in the form

$$F(x_1, x_2, \ldots, x_n) = \sum_{i_1 = 0}^{1} \sum_{i_2 = 0}^{1} \cdots \sum_{i_n = 0}^{1} F(i_1, i_2, \ldots, i_n) . P_j$$

where

$$P_j = P_{i_1 i_2, \ldots, i_n} = x_1^{i_1} . x_2^{i_2} \ldots \ldots x_n^{i_n}$$

and

$$x_k^{i_k} \text{ means } \bar{x}_k \text{ if } i_k = 0$$

and

$$x_k \text{ if } i_k = 1$$

(The symbol $\sum_{j=1}^{n} y_j$ is used for a Boolean sum $y_1 + y_2 + \ldots + y_n$.)

9. Write out in full the result of the previous question for the particular case
   of three variables $x_1 = x$, $x_2 = y$, $x_3 = z$.
   Hence determine the special conjunctive normal form for the Boolean
   function that takes the value 1 if any two or more of the three variables x, y, z
   are 1 and takes the value 0 otherwise.

10. Write out the special disjunctive normal form for each of the following
    functions:
    (a) $f(x, y, z) = x + \bar{y}$
    (b) $f(w, x, y, z) = w . x . y . z$
    (c) $f(x, y, z) = (x + y) . (\bar{x} + z)$
    (d) $f(x, y, z) = \overline{x . (\bar{y} + z) + (\overline{x . y})}$

11. Obtain the special conjunctive normal form for each of the functions of Question 10.

12. Show that a Boolean algebra cannot consist of exactly three distinct elements.

Chapter 4
The Algebra of Propositions

Introduction

The laws of logic have been studied for many centuries, since the time of Aristotle. During the last hundred years, from the time of George Boole, powerful mathematical techniques have been introduced and the study of logic has become a branch of mathematics known as 'symbolic logic' or 'mathematical logic'. This chapter deals with a small portion of the subject known as the 'algebra of propositions'. It will be treated here on an informal basis as was done with the algebra of classes. A formal approach is also possible, in which all the theorems are obtained from a set of axioms and given methods of derivation. This formal approach is generally called the 'calculus of propositions'.

There is a very close connection between the algebra of propositions, Boolean algebra, and logic circuits. The techniques derived in the algebra of propositions will prove fundamental to the design and analysis of logic or switching circuits.

Propositions, Propositional Variables, and Propositional Connectives

As in the algebra of sets or elementary geometry, it is necessary to introduce certain undefined terms. These are *proposition, true,* and *false*. To be able to apply the algebra to actual statements, the intuitive meanings of these terms should be applicable to the formal undefined terms. Thus *true* and *false* are attributes of the respective statements,

'London is the capital of England'

and

'Three is an even number'

This type of statement, about which it is possible to say that it is either true or false but not both, is called a *proposition*. A proposition therefore can take the *truth value* either true or false. No other value is permitted and the algebra of propositions thus refers to a two-valued logic.

It is not necessary to know with certainty which of the two values true or false holds for a given statement to be a proposition. Thus

'There is life on Mars'

is a proposition since it must be true or false even though it may not be known at the present time. It is very necessary that the truth value be unambiguous. There are several well-known paradoxes to illustrate this.

(a) Epimenides, the Cretan, made the statement

'All Cretans always lie'

(b)   'The sentence quoted in this line is false'

Now, in both cases, if the statement is assumed to be true then from the content of the statement it must be false and vice versa. These two statements may therefore not be considered as propositions.

It is convenient to represent propositions by letters x, y, z, etc. If it represents an arbitrary proposition, it is known as a *propositional variable.*

From given simple propositions, further compound propositions can be formed by means of *propositional connectives.* The simplest is a singulary operator that forms the *negation* of a proposition. This is written as $\bar{x}$ for the negation of the proposition x. It may be translated by the proposition

'It is false that x'

Thus, if x is the proposition

'Mathematics is an easy subject'

then the negation $\bar{x}$ is the proposition

'It is false that mathematics is an easy subject'

Equivalent ways of saying this are

'Mathematics is not an easy subject'

or

'Mathematics is a difficult subject'

The negation of a proposition always has the opposite truth value to the proposition itself. Thus, if x is true, $\bar{x}$ is false and, if x is false, $\bar{x}$ is true. This may be shown in a *truth table,* as in Table 4.1.

Table 4.1. Truth table for negation

x	$\bar{x}$
F	T
T	F

The two simplest binary connectives are those of conjunction and disjunction (or alternation). Let y be the proposition

'4 is a prime number'

then the conjunction of x and y, written $x \& y$ (read as 'x and y') is the proposition

'Mathematics is an easy subject and 4 is a prime number'

The compound proposition is true if and only if *both* of the simple propositions x and y are true.

The truth table for this is shown in Table 4.2.

Table 4.2. Truth table for conjunction

x	y	$x \& y$
F	F	F
F	T	F
T	F	F
T	T	T

The disjunction of the two propositions x and y is $x \lor y$ (read as 'x or y'):

'Mathematics is an easy subject or 4 is a prime number'

This is true if x or y or both are true. This will be the normal meaning of the 'or' connective, that is *inclusive* disjunction. There is another connective 'exclusive or' which is true if x or y but not both are true. This operation is also known as 'non-equivalence' and is written as $x \not\equiv y$.

Table 4.3. Truth table for disjunction   Table 4.4. Truth table for non-equivalence

x	y	$x \lor y$
F	F	F
F	T	T
T	F	T
T	T	T

x	y	$x \not\equiv y$
F	F	F
F	T	T
T	F	T
T	T	F

Truth Tables

Truth tables were used in the previous section to illustrate the connectives $^-$, &, V but they play a very important role in the algebra of propositions. The concept of 'equality' has not yet been defined for this algebra.

In Chapter 3 it was shown that a Boolean function was uniquely determined by the values taken when each of the variables was allotted all possible combinations of 0 and 1. This suggests a method for propositions using truth values F and T for false and true respectively. The truth table for any propositional function f of the variables $x_1, x_2, \ldots, x_n$ gives the truth value of the function for every possible way of assigning the value F or T to the variables $x_1, x_2, \ldots, x_n$. It is conventional to equate the value F with 0 and T with 1. For a function with n variables, there are 2^n ways of assigning the values 0 and 1 to the variables and there must be 2^n rows in the truth table. To ensure that none are missed, write the successive rows so that if they are read as binary numbers they represent 0 to $2^n - 1$. Thus the truth table for the function $(x \& y) \& z$, which will yield the value 1 only if $(x \& y)$ is 1 and z is 1, i.e. if x and y and z are all 1, is as shown in Table 4.5.

Table 4.5. Truth table for $(x \& y) \& z$

Row No.	x	y	z	$(x \& y) \& z$
0	0	0	0	0
1	0	0	1	0
2	0	1	0	0
3	0	1	1	0
4	1	0	0	0
5	1	0	1	0
6	1	1	0	0
7	1	1	1	1

The definition of equality for two propositional functions is that they should have an identical set of truth values in their truth tables. If the truth values differ in any row, the functions are not equal. The word for this type of equality in logic is *equiveridic*. Thus, if two propositional functions f, g take the same set of truth values, f is equiveridic to g. This will be written as $f = g$. If it is particularly important to distinguish between equiveridic and equals, then the notation f eq. g can be used.

If the truth table for a function f contains only 1s in the final column, the proposition is called a *tautology*. A tautology is therefore a compound proposition that is always true regardless of the truth or falsity of its constituent variables. Any substitution of truth values for the variables yields the result

$$f = 1$$

It will sometimes be convenient to denote a tautology by the symbol I as the name of the universal element of the algebra of propositions corresponding to the universal set in the algebra of classes.

As a simple example of a tautology, consider the combined proposition

'Mathematics is easy or mathematics is not easy'

i.e. a proposition of the form $x \vee \bar{x}$.

Table 4.6. The tautology $x \vee \bar{x}$

x	$\bar{x}$	$x \vee \bar{x}$
0	1	1
1	0	1

The truth table for $x \vee \bar{x}$ is shown in Table 4.6. This truth table is true for *any* proposition x. Thus the function

$$x \vee \bar{x} = 1$$

The opposite of a tautology is known as a *contradiction*. This is a compound proposition that always takes the truth value 0 for all values of the primitive variables. Its truth table consists of all zeros. A simple example of a contradiction is $x \And \bar{x}$. A proposition f is a tautology if and only if $\bar{f}$ is a contradiction.

To find the truth value for more complicated functions, it is advisable to build up the values in easy steps using the truth tables for the operations $^-$, $\And$, $\vee$. Table 4.7 shows the steps in finding the truth value of $f = x \vee \overline{y \And z}$.

Table 4.7. Truth table of $x \vee \overline{y \And z}$

x	y	z	$y \And z$	$\overline{y \And z}$	$x \vee \overline{y \And z}$	$f = x \vee \overline{y \And z}$
0	0	0	0	1	1	0
0	0	1	0	1	1	0
0	1	0	0	1	1	0
0	1	1	1	0	0	1
1	0	0	0	1	1	0
1	0	1	0	1	1	0
1	1	0	0	1	1	0
1	1	1	1	0	1	0

To avoid ambiguities, parentheses must be inserted, where necessary in an expression to indicate the order of evaluation. Additional parentheses may, of course, be inserted in the interests of clarity. As in Boolean algebra, there is a conventional order of priorities. The order of precedence in descending order is

(i)   negation $^-$
(ii) conjunction $\And$
(iii) disjunction $\vee$

In the above function therefore, with all parentheses shown,

$$f = \overline{(x \vee (\overline{y \And z}))}$$

The expression $x \lor y \,\&\, z$ is understood to represent $x \lor (y \,\&\, z)$. If $(x \lor y) \,\&\, z$ is required, the parentheses *must* be present. Thus the expression

$$w \,\&\, \overline{x \lor y} \lor y \,\&\, z$$

would represent

$$(w \,\&\, (\overline{x \lor y})) \lor (y \,\&\, z)$$

The truth table for $\overline{x} \,\&\, y \,\&\, z$ is given in Table 4.8. This is seen to give the identical result to Table 4.7. Thus equiveridicity of the two functions has been shown:

$$\overline{x \lor y \,\&\, z} = \overline{x} \,\&\, y \,\&\, z$$

Table 4.8. Truth table for $\overline{x} \,\&\, y \,\&\, z$

x	y	z	$\overline{x}$	$\overline{x} \,\&\, y$	$(\overline{x} \,\&\, y) \,\&\, z$
0	0	0	1	0	0
0	0	1	1	0	0
0	1	0	1	1	0
0	1	1	1	1	1
1	0	0	0	0	0
1	0	1	0	0	0
1	1	0	0	0	0
1	1	1	0	0	0

Connection with Boolean Algebra

It is now possible to show that the algebra of propositions is a Boolean algebra.
The two binary operations are
   (i) conjunction 'AND' (&) corresponding to the Boolean product
   (ii) disjunction 'OR' ($\lor$) corresponding to the Boolean sum

Equiveridic corresponds to equality
(F) 'False' corresponds to the zero element 0
(T) 'True' corresponds to the unit element 1

It is now necessary to show that the four axioms of a Boolean algebra are satisfied.
   (i) The operations & and $\lor$ are commutative. This follows immediately from the definitions.
   (ii) Identity elements are 0 and 1.

$$x \lor 0 = x$$

since the disjunction is true precisely when x is true.

$$x \,\&\, 1 = x$$

since the conjunction is true only when x is true.

(iii) Each operation is distributive over the other, i.e.

$$x \mathbin{\&} (y \vee z) = (x \mathbin{\&} y) \vee (x \mathbin{\&} z)$$
$$x \vee (y \mathbin{\&} z) = (x \vee y) \mathbin{\&} (x \vee z)$$

Both these equations can easily be verified by writing out the truth tables.

(iv) An inverse element exists. The inverse of a proposition x is the negation $\bar{x}$.

$$x \mathbin{\&} \bar{x} = 0$$
$$x \vee \bar{x} = 1$$

Since all four axioms hold true for the algebra of propositions, it may be treated as a Boolean algebra and all the theorems and techniques derived in the previous chapter may be applied. The conjunction operation & corresponds to the Boolean product . and disjunction $\vee$ corresponds to the Boolean sum +. For example, De Morgan's law (Theorem 3.4)

$$\overline{x \cdot y} = \bar{x} + \bar{y}$$

can be translated to

$$\overline{x \mathbin{\&} y} = \bar{x} \vee \bar{y}$$

which may be paraphrased as

'It is false that both x and y are true'
has the same truth values as 'x is false
or y is false (or both)'

Other Propositional Connectives

Three connectives have been defined so far. There are several more, but, before a complete list is given, there is one of particular importance which should be discussed in detail.

Consider the two propositions

x denoting 'It is raining'
y denoting 'The street is wet'

and draw up a truth table to represent the function that determines whether or not the two conditions coexist in the physical world. Examining the four possibilities, it is clear that the combination of

'It is true that it is raining'

with

'It is false that the street is wet'

is not possible physically, while the other three are all possible results. This gives the truth table of Table 4.9.

The operation defined by the above truth table is known as *implication* and written $x \rightarrow y$ (x implies y). In propositions it arises in the form:

'If . . . then . . .'

Thus 'If it is raining, then the street is wet'. The term x is often known as the antecedent of the conditional $x \rightarrow y$ and y as the consequent.

Table 4.9. Truth table for implication

x	y	f
0	0	1
0	1	1
1	0	0
1	1	1

From the truth table of Table 4.9 it can be seen that the truth values are the same as $\bar{x} \vee y$, so that

$$x \rightarrow y = \bar{x} \vee y$$

Thus 'If x then y' or 'x implies y' means 'Either x is false or y is true' and no further meaning should be read into the word 'implies' in mathematical logic This is also the normal meaning of the conditional statement 'If . . . then . . .' in mathematics. For example, consider the statement

'If a number N is an even positive integer, then its square N^2 is also an even positive integer'

This would be considered as a true statement and examples in which $N = -2$ or $N = +3$ would not be considered as counterexamples in any investigation. Thus the first and second lines in Table 4.9 in which x is false still lead to a true combined statement.

In everyday English, a different interpretation is sometimes given to conditional statements. Thus the statement

'If the sun shines, I will go swimming'

would be understood by many people to mean that

'If the sun does not shine, I will not go swimming'

whereas logically no such meaning should be applied to the statement. There-fore, in the algebra of propositions, the only line in the truth table that gives a false result is when a true value for the antecedent x implies a false value for the consequent y as in the statement

'If London is in England, then $2 + 2 = 5$'

Two statements which would be considered as true in mathematical logic are

'If London is in Scotland, then $2 + 2 = 4$'

and

'If the moon were made of green cheese, then pigs could fly'

The type of implication used in the above examples in which a compound statement $x \to y$ is formed from two simple propositions x, y by using the $\to$ connective, is known as *material* implication. This is to distinguish it from an alternative usage known as *logical* implication. Consider the two propositions

x: 'If it is raining, then the street is wet'

and

y: 'It is raining'

Then from these two propositions can be deduced the fact

z: 'The street is wet'

Whenever a statement or set of statements cannot be true without a further statement *necessarily* being true, then the former logically implies the latter. So that in the above example

$x \& y$ logically implies z

Logical implication may be defined generally by the following paragraph.

Consider two propositional functions $f(x, y, z, \ldots)$ and $g(x, y, z, \ldots)$. From these it is possible to form the compound proposition $f \to g$. If this proposition is a tautology, i.e. the truth table contains only 1s, this means that the implication is true regardless of the individual truth values of the simple propositions $x, y, z, \ldots$ that make up the functions. In this case, f is said to *imply logically* g, which is sometimes written as

$$f \Rightarrow g$$

Another way of stating this result is that g is a *logical consequence* of f. Note carefully the difference in meaning between the two statements

(a) $f \to g$
(b) $f \Rightarrow g$

(a) is a compound proposition formed with the connective $\to$, and its truth table may contain any combination of 0s and 1s. It could equally be written as $\bar{f} \vee g$.

(b) is a relation between two propositions f and g which has the meaning that $f \to g$ or $\bar{f} \vee g$ is a tautology. This relation satisfies the reflexive, antisymmetric, and transitive laws as for example does the partial order relation ($\leqslant$) of .

Boolean algebra discussed in Chapter 3, the inclusion relation ($\subset$) of the algebra of classes or the inequality $\geq$ in conventional algebra.

Some examples of logical implication are given by the following:

(i) $(x \& y) \rightarrow (x \vee y)$

(ii) $[(x \rightarrow y) \& (y \rightarrow z)] \rightarrow (x \rightarrow z)$

Each of these compound propositions is a tautology. Hence

(i) $x \& y \Rightarrow x \vee y$

(ii) $(x \rightarrow y) \& (y \rightarrow z) \Rightarrow x \rightarrow z$

It should be noted that $\rightarrow$ is not a commutative operation like & and $\vee$.

$$x \rightarrow y = \bar{x} \vee y$$
$$y \rightarrow x = \bar{y} \vee x = x \vee \bar{y}$$

'$y \rightarrow x$' is known as the *converse* of '$x \rightarrow y$'. Nor is it associative, so that parentheses must be used. Truth tables for $(x \rightarrow y) \rightarrow z$ and $x \rightarrow (y \rightarrow z)$ are shown in Table 4.10.

Table 4.10. Truth tables of $(x \rightarrow y) \rightarrow z$ and $x \rightarrow (y \rightarrow z)$

x	y	z	$x \rightarrow y$	$y \rightarrow z$	$(x \rightarrow y) \rightarrow z$	$x \rightarrow (y \rightarrow z)$
0	0	0	1	1	0	1
0	0	1	1	1	1	1
0	1	0	1	0	0	1
0	1	1	1	1	1	1
1	0	0	0	1	1	1
1	0	1	0	1	1	1
1	1	0	1	0	0	0
1	1	1	1	1	1	1

Another connective that occurs frequently in mathematics is that of *equivalence*, i.e. 'if and only if'. The two common notations for this are

$$x \equiv y$$

or

$$x \leftrightarrow y$$

This may be defined by

$$x \equiv y = (x \rightarrow y) \& (y \rightarrow x)$$

or

$$x \equiv y = (x \& y) \vee (\bar{x} \& \bar{y})$$

or

$$x \equiv y = \overline{x \not\equiv y}$$

The truth table is shown in Table 4.11.

The ideas of 'necessary and sufficient' conditions that are quoted so freely in mathematical theorems can be defined quite precisely by means of the connectives $\rightarrow$ and $\equiv$. A condition c is necessary for a result r if

$$r \rightarrow c$$

is a true proposition. A condition c is sufficient for a result r if

$$c \rightarrow r$$

is true.

Table 4.11. Truth table of equivalence

x	y	$x \equiv y$
0	0	1
0	1	0
1	0	0
1	1	1

If c is a necessary and sufficient condition, then

$$r \rightarrow c \quad \text{and} \quad c \rightarrow r$$

are both true, that is to say

$$c \equiv r$$

is true.

Instead of picking out arbitrary connectives to discuss, consider now all the possible propositional functions that can be formed from two variables x, y.

Table 4.12. Propositional functions of two variables

y	z	f_{15}	f_{14}	f_{13}	f_{12}	f_{11}	f_{10}	f_9	f_8	f_7	f_6	f_5	f_4	f_3	f_2	f_1	f_0
0	0	1	1	1	1	1	1	1	1	0	0	0	0	0	0	0	0
0	1	1	1	1	1	0	0	0	0	1	1	1	1	0	0	0	0
1	0	1	1	0	0	1	1	0	0	1	1	0	0	1	1	0	0
1	1	1	0	1	0	1	0	1	0	1	0	1	0	1	0	1	0

There are 16 different sets of truth values as shown in Table 4.12. If the diagram is rotated $90°$ anticlockwise, it will be seen that each of the functions is represented by four binary digits giving the number of the function in binary. Thus, for example,

$$f_1 \text{ has truth values } 0001$$
$$f_{10} \text{ has truth values } 1010$$

A full listing of the 16 functions with the names of the connectives and the symbols to represent them is given in Table 4.13.

Table 4.13. Propositional connectives

Function number	Binary number	Name of connective	Meaning of connective	Symbol	Equivalent symbol	Other common symbols
0	0000	null element	always false	0		$\emptyset, o$
1	0001	conjunction	y and z	$y \, \& \, z$	$y \cdot z$	$\cap, \wedge, \times$
2	0010	non-implication	y but not z	$\overline{y \to z}$		$\wedge$
3	0011	(primitive element)	$\ldots$	y		
4	0100	non-inclusion	z but not y	$\overline{y \leftarrow z}$		$\vee$
5	0101	(primitive element)	$\ldots$	z		
6	0110	non-equivalence	y exclusive or z	$y \not\equiv z$	$y + z$	$\oplus \vee$
7	0111	disjunction	y or z	$y \vee z$	$y \to z$	$\cup$
8	1000	joint denial	y nor z	$\overline{y \vee z}$		$\overline{\cup}, \nabla$
9	1001	equivalence	y if and only if z	$y \equiv z$		$\leftrightarrow$
10	1010	negation	not z	$\bar{z}$		$z', \sim z, \neg z$
11	1011	inclusion	if z then y	$y \leftarrow z$		$\geqslant, \cup$
12	1100	negation	not y	$\bar{y}$		$y', \sim y, \neg y$
13	1101	implication	if y then z	$y \to z$		$\leqslant, \cap$
14	1110	non-conjunction	y and z	$\overline{y \, \& \, z}$	$y \mid z$	$\bar{\cap}, \bar{\wedge}, \bar{\&}$
15	1111	universal element	always true, tautology	1		I, i

As will be seen from the table, there are many alternative symbols used for the various propositional connectives. There is no single set that may be regarded as a standard, and the reader must be prepared to meet any of them in different textbooks.

There are two new operations appearing in the table that will be of particular importance later in the book. The first is joint denial $\overline{y \lor z}$ which is NOT (y OR z) and may be considered as a single binary operation $y \overline{\lor} z$, read as y NOR z, with the meaning 'neither y nor z'. It is frequently convenient to have a single symbol for the NOR operation and the most usual one is the dagger operator, $y \downarrow z$. The other important operation is non-conjunction $\overline{y \And z}$, usually referred to as NAND, i.e. NOT (y AND z). The single operator equivalent is the *Sheffer stroke*, $y|z$.

Synthesis of Functions from the Truth Table

Truth tables so far have been used to define operations and to prove equality between functions, that is essentially an analysis of a given function. The opposite procedure is also possible. Given a truth table definition it is a simple matter to write down the special disjunctive or conjunctive normal forms of the function. Consider the truth table of Table 4.14. The function takes the

Table 4.14. Truth table definition of a function f

x	y	z	f
0	0	0	1
0	0	1	0
0	1	0	0
0	1	1	1
1	0	0	1
1	0	1	0
1	1	0	0
1	1	1	1

value 1 or true in rows 0, 3, 4, and 7 of the truth table. Thus the value 1 is required when either row 0 OR row 3 OR row 4 OR row 7 is true. Row 0 is true for $\overline{x} \And \overline{y} \And \overline{z}$ true (i.e. $x = 0$, $y = 0$, $z = 0$); row 3 for $\overline{x} \And y \And z$ true, etc. Thus

$$f = \overline{x} \And \overline{y} \And \overline{z} \lor \overline{x} \And y \And z \lor x \And \overline{y} \And \overline{z} \lor x \And y \And z$$

which is the propositional equivalent of the Boolean special disjunctive normal form of the function,

$$(f = \overline{x} . \overline{y} . \overline{z} + \overline{x} . y . z + x . \overline{y} . \overline{z} + x . y . z)$$

The conjunctive normal form may be obtained from this by using De Morgan's laws. Thus

$$f = \overline{\overline{x \,\&\, \overline{y} \,\&\, \overline{z}} \vee \overline{\overline{x} \,\&\, y \,\&\, z} \vee \overline{x \,\&\, \overline{y} \,\&\, \overline{z}} \vee \overline{x \,\&\, y \,\&\, z}}$$
$$= (x \vee y \vee z) \,\&\, (x \vee \overline{y} \vee \overline{z}) \,\&\, (\overline{x} \vee y \vee z) \,\&\, (\overline{x} \vee \overline{y} \vee \overline{z})$$

The final negation can be obtained by writing out the terms that are missing from the complete normal form, i.e.

$$f = (x \vee y \vee \overline{z}) \,\&\, (x \vee \overline{y} \vee z) \,\&\, (\overline{x} \vee y \vee \overline{z}) \,\&\, (\overline{x} \vee \overline{y} \vee z)$$

This expression can equally well be obtained direct from the truth table of Table 4.14 by applying the principle of duality direct. To obtain the special conjunctive normal form, pick out the rows of the truth table containing zeros, i.e. rows 1, 2, 5, 6. Row 1 takes the value 0 if $x \vee y \vee \overline{z} = 0$ (i.e. $x = 0, y = 0$, $z = 1$).

Functionally complete sets of operations

It has been shown that the algebra of propositions is a Boolean algebra. In a Boolean algebra, however, there are only three operations $^{-}$, ., +, corresponding to $^{-}$, &, $\vee$ respectively. Can the other connectives introduced in the preceding section be expressed in terms of these operations? Several of them have already been defined in these terms, e.g.

$$x \rightarrow y = \overline{x} \vee y$$
$$x \equiv y = (x \,\&\, y) \vee (\overline{x} \,\&\, \overline{y})$$

Now each of the propositional functions is defined by its set of truth values in the truth table and every truth table has a unique special disjunctive normal form using only the operations $^{-}$, &, $\vee$. Thus each of the functions defined in Table 4.12 may be written in terms of these three operations.

Is it possible to obtain any new functions by investigating functions of three or more propositional variables? The set of truth values of any function that is defined will be given by its truth table and thus can be converted into disjunctive normal form and therefore expressed by the set of operations $^{-}$, &, $\vee$. These three operations will therefore determine all propositional functions.

A set of operations in which all propositional functions may be expressed is called functionally complete. Thus the set $^{-}$, &, $\vee$ is functionally complete.

Since all propositional functions can be expressed in terms of $^{-}$, &, $\vee$ and also form a Boolean algebra in which conjunction (&) and disjunction ($\vee$) represent the Boolean product (.) and Boolean sum (+) respectively, it is often convenient to write propositional functions in terms of the Boolean symbols. It is then easy to apply any theorems of Boolean algebra and methods of simplification. These are also the symbols (. +) in most common use for logic circuitry. Thus equivalence can be written in disjunctive normal form as

$$x \equiv y = x \,\&\, y \lor \bar{x} \,\&\, \bar{y}$$
$$= x \,.\, y + \bar{x} \,.\, \bar{y}$$
$$= xy + \bar{x}\bar{y}$$

where the product sign (.) has been omitted in the last line and simple juxtaposition of variable used to denote a product as in ordinary multiplication. The sign will be omitted in this way if no confusion is likely to arise. Of course the correspondence between the algebra of propositions and Boolean algebra works in both directions. Truth table methods can be used for Boolean algebra, for example, to express a Boolean function in special normal form.

Still simpler sets of functionally complete operators can be found. By De Morgan's laws, conjunction may always be expressed in terms of negation and disjunction

$$\overline{x \,\&\, y} = \bar{x} \lor \bar{y}$$

Therefore

$$x \,\&\, y = \overline{\bar{x} \lor \bar{y}}$$

so that all conjunctions may be removed from the disjunctive normal form leaving an expression entirely in $^{-}$ and $\lor$. Thus, $^{-}$, $\lor$ is a functionally complete set. In a similar manner, disjunction can be removed instead, giving $^{-}$, $\&$ as functionally complete.

It is also possible to express all the functions in terms of negation and implication so that $^{-}$, $\rightarrow$ is functionally complete. As an example

$$x \,\&\, y = \overline{x \rightarrow \bar{y}}$$
$$\cdot \quad x \lor y = \bar{x} \rightarrow y$$

so that by converting $\&$, $\lor$ with the above formulae, expressions in $^{-}$, $\rightarrow$ only are obtained. Expressions in this form are much more difficult to manipulate but are frequently used in mathematical treatments of the calculus of propositions.

The NOR, NAND functions are particularly interesting as either of them forms a functionally complete set in itself. Any propositional function can be expressed in terms of $\downarrow$ only or $|$ only.

Consider the NOR operation and note first that

$$x \downarrow x = \bar{x} \lor \bar{x} = \bar{x}$$

so that negation may be removed. An alternative expression is

$$\bar{x} = x \downarrow 0$$

Also

$$x \lor y = \overline{\overline{x \lor y}} = \overline{x \downarrow y}$$

so that eliminating NOT gives

$$x \lor y = (x \downarrow y) \downarrow (x \downarrow y)$$

Some care must be taken in manipulating these symbols and in the use of parentheses. The NOR and NAND connectives are not associative.

$$(x \downarrow y) \downarrow z = \overline{(x \lor y)} \lor z$$
$$= \overline{x \lor y} \,\&\, \overline{z}$$
$$= (x \lor y) \,\&\, \overline{z}$$

while

$$x \downarrow (y \downarrow z) = \overline{x} \,\&\, (y \lor z)$$

The expression $x \downarrow y \downarrow z$ may also be defined and is different from either of the previous expressions. $x \downarrow y \downarrow z$ is defined to mean $\overline{x \lor y \lor z}$ and is equiveridic to $\overline{x} \,\&\, \overline{y} \,\&\, \overline{z}$. It is true only when not one of the variables is true, i.e. when all variables are false.

Table 4.15. Propositional connectives expressed in terms of functionally complete sets

	Binary number	NOT/AND	NOT/OR	NOT/IMPLIES
O	0 0 0 0	$y \,\&\, \overline{y}$	$\overline{y \lor \overline{y}}$	$\overline{y \rightarrow y}$
$y \,\&\, z$	0 0 0 1	$y \,\&\, z$	$\overline{\overline{y} \lor \overline{z}}$	$\overline{y \rightarrow \overline{z}}$
$\overline{y \rightarrow z}$	0 0 1 0	$y \,\&\, \overline{z}$	$\overline{\overline{y} \lor z}$	$\overline{y \rightarrow z}$
y	0 0 1 1	y	y	y
$\overline{y \leftarrow z}$	0 1 0 0	$\overline{y} \,\&\, z$	$\overline{y \lor \overline{z}}$	$\overline{\overline{y} \rightarrow \overline{z}}$
z	0 1 0 1	z	z	z
$y \not\equiv z$	0 1 1 0	$\overline{y \,\&\, z \,\&\, \overline{y} \,\&\, \overline{z}}$	$\overline{\overline{y} \lor z} \lor \overline{y \lor \overline{z}}$	$(y \rightarrow z) \rightarrow \overline{\overline{y} \rightarrow \overline{z}}$
$y \lor z$	0 1 1 1	$\overline{\overline{y} \,\&\, \overline{z}}$	$y \lor z$	$\overline{y} \rightarrow z$
$\overline{y \lor z}$	1 0 0 0	$\overline{y} \,\&\, \overline{z}$	$\overline{y \lor z}$	$\overline{\overline{y} \rightarrow z}$
$y \equiv z$	1 0 0 1	$\overline{\overline{y} \,\&\, z \,\&\, \overline{y \,\&\, \overline{z}}}$	$\overline{y \lor z} \lor \overline{\overline{y} \lor \overline{z}}$	$\overline{(y \rightarrow z) \rightarrow \overline{\overline{y} \rightarrow \overline{z}}}$
$\overline{z}$	1 0 1 0	$\overline{z}$	$\overline{z}$	$\overline{z}$
$y \leftarrow z$	1 0 1 1	$\overline{\overline{y} \,\&\, z}$	$y \lor \overline{z}$	$\overline{y} \rightarrow \overline{z}$
$\overline{y}$	1 1 0 0	$\overline{y}$	$\overline{y}$	$\overline{y}$
$y \rightarrow z$	1 1 0 1	$\overline{y \,\&\, \overline{z}}$	$\overline{y} \lor z$	$y \rightarrow z$
$\overline{y \,\&\, z}$	1 1 1 0	$\overline{y \,\&\, z}$	$\overline{y} \lor \overline{z}$	$y \rightarrow \overline{z}$
I	1 1 1 1	$\overline{y \,\&\, \overline{y}}$	$y \lor \overline{y}$	$y \rightarrow y$

N.B. The last column can be simplified if the equiveridic forms $y \leftarrow z = z \rightarrow y$ are used instead of $y \leftarrow z = \overline{y} \rightarrow \overline{z}$.

Expressions for the 16 possible propositional functions in terms of the functionally complete sets NOT, AND; NOT, OR; NOT, IMPLIES are given in Table 4.15. The corresponding expression for NAND only can easily be obtained from these by the substitutions

$$\bar{x} = x \mid x \text{ or } x \mid 1$$

$$x \& y = (x \mid y) \mid (x \mid y) \text{ or } (x \mid y) \mid 1$$

and, those for NOR only, by the substitutions given earlier.

Multi-valued Logic

The algebra of propositions that has been considered has only had two possible truth values, false and true. It is possible to set up systems of logic with more possible values. The truth values 0 to 3 could be interpreted as false, implausible, plausible, true.

In the general case there would be n truth values $1, 2, \ldots, n$ and the operations $^-$, &, V can be defined by the truth tables shown in Tables 4.16, 4.17, 4.18. The value n plays the role of false in the two-valued system.

Table 4.16. Truth table for V operation   Table 4.17. Truth table for & operation

x	1	2	3 . . . n
y			
1	1	1	1 . . . 1
2	1	2	2 . . . 2
3	1	2	3 . . . 3
.	.		.
.	.		.
.	.		.
n	1	2	3 . . . n

x	1	2	3 . . . n
y			
1	1	2	3 . . . n
2	2	2	3 . . . n
3	3	3	3 . . . n
.	.		.
.	.		.
.	.		.
n	n	n	n . . . n

Table 4.18. Truth table for $^-$ operation

x	$\bar{x}$
1	2
2	3
3	4
.	.
.	.
.	.
$n-1$	n
n	1

Using these truth table definitions for the basic operations, an abstract algebra known as a Post algebra may be set up. Multi-valued logics have not as yet found much practical application.

Logical Arguments

As was mentioned at the beginning of this chapter, the laws of logic were studied by Aristotle. A classical example of the type of logical argument known as a *syllogism* that was considered by Aristotle is as follows:

(a) All men are mortal
(b) Socrates is a man
_____
(c) Socrates is mortal

In this example, given the two propositions (a) and (b) which are known as the *premises* (and are assumed to be true), is it possible to assert proposition (c) as a valid *conclusion*? This process is known as *deduction* and it is very easy to show by the algebra of sets that the deduction is correct *in this case*.

In Figure 4.1 let the universal set *I* be the set of all living things.

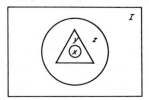

Figure 4.1

Let   *y* denote the set of all men
                                 z be the set of all mortals
                                 x be the set whose only member is Socrates

The two premises state that

(a) $y \subset z$
(b) $x \subset y$

Therefore by the *transitive law* of set inclusion

$$x \subset z$$

or 'Socrates is mortal'.

This syllogism is valid for all schemata of the same form, i.e.

All *y* are *z*
All *x* are *y*
_____
All *x* are *z*

Thus having shown that this is a valid schema, starting from the two premises

(a) All English people are British
(b) All Londoners are English

a valid conclusion is 'All Londoners are British'. The conclusion is true independently of the particular statements used. The conclusion is *logically implied* by the premises, or

$$(y \subset z) \,\&\, (x \subset y) \Rightarrow (x \subset z)$$

This example may also be expressed in the notation of the algebra of propositions. In order to achieve this, each premise can be interpreted as a compound proposition in the form of an implication. Thus 'All y are z' may be written as '$y \rightarrow z$' since $y \rightarrow z = 1$ (i.e. true) if the original premise is true, because under this premise the combination $y = 1$, $z = 0$ does not occur. This syllogism may therefore be written in the form

$$\frac{\begin{array}{c} y \rightarrow z \\ x \rightarrow y \end{array}}{x \rightarrow z}$$

This is the most important syllogism and is often referred to as *the law of syllogism.*

A logical argument is valid if the conjunction of the premises $P_1, P_2, \ldots, P_n$ logically implies the conclusion C, i.e.

$$P_1 \,\&\, P_2 \,\&\, \ldots \,\&\, P_n \Rightarrow C$$

The validity of an argument of this form may therefore be proved by showing that

$$P_1 \,\&\, P_2 \,\&\, \ldots \,\&\, P_n \rightarrow C$$

is a tautology, either by means of a truth table or by algebraic simplification. Thus the above law of syllogism follows from the tautology

$$[(y \rightarrow z) \,\&\, (x \rightarrow y)] \rightarrow (x \rightarrow z)$$

which was given earlier in the chapter.

In more general arguments the validity of the deduction depends not only on the truth values of the individual propositions but also on the internal structure of the statements. Thus the propositions involved depend on the use of such words as 'all', 'some', and 'any'. Further, a proposition is considered to consist of a subject and a predicate.

A special notation is used to denote that a property holds for all members of a set. If the property of an individual y being a mortal is denoted by $M(y)$, then the proposition 'For all x, x is a mortal' is written as $(x)M(x)$. The symbol '(x)' is known as a *universal quantifier,* and the predicate M is 'is a mortal'. There is another quantifier known as an *existential quantifier* which denotes that there exists at least one object that has the property P. This is written as $(\exists z)P(z)$. The study of logical systems obtained by including quantifiers as well as the simple algebra of propositions is generally known as *predicate calculus*. Unfortunately there is no decision procedure in the predicate calculus corresponding to the truth table method of the algebra of propositions. Thus there is no simple method to determine whether or not any given formula is valid.

The predicate calculus is much too large a subject to be considered further here. However, as in the previous example, simple results involving the use of the universal quantifier may be obtained from the algebra of classes and Venn diagrams. The syllogisms of Aristotle are particular forms of deduction in which the two premises contain a common or middle term (*y* in the previous example) and the conclusion is a proposition from which the middle term is absent. The validity of the syllogisms, which had to be learned by heart by the Ancient Greeks using a system of mnemonics, can very easily be determined by Venn diagrams.

As a further example of a deduction, consider the premises

    (i)   All students are poor
    (ii) Film stars are famous
    (iii) Richard is a student
    (iv) No famous person is poor

What conclusion can be reached? There is of course more than one possible conclusion, e.g. from (i) and (iii) 'Richard is poor'. But this does not use all the premises.

    Let
        the class of all people be the universal class I
        the class of students be S
        the class of poor people be P
        the class of film stars be F
        the class of famous people be M
        the unit class consisting of Richard be R

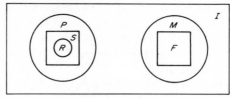

Figure 4.2

The Venn diagram of Figure 4.2 may then be drawn by noting
    (a) class P and M are disjoint by (iv)
    (b) $S \subset P$
    (c) $R \subset S$
    (d) $F \subset M$

Using all the premises, it can be seen from the Venn diagram that R and F are disjoint, i.e. 'Richard is not a film star'.

Now, as a warning, a syllogism that is *not* valid is given. It is known as the law of the undistributed middle. This is of the type

$$\begin{array}{l} \text{All } x \text{ are } y \\ \underline{\text{All } z \text{ are } y} \\ \text{All } x \text{ are } z \end{array}$$

e.g.   All differentiable functions are continuous
All polynomials are continuous
———————————————————
All polynomials are differentiable

Let
the universal class be the class of all mathematical functions
the class of continuous functions be C
the class of differentiable functions be D
the class of polynomials be P

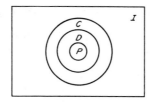

Figure 4.3

The Venn diagram in Figure 4.3 may then be drawn from this information. From this figure, the conclusion $P \subset D$, i.e. 'All polynomials are differentiable' would appear to be valid. In fact the conclusion, although it happens to be correct in this instance, is not derivable from the premises, since the Venn diagram of Figure 4.4 also satisfies the premises. In this case, however, P is not included in D. An argument that is not valid is known as a *fallacy*.

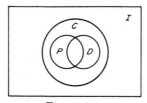

Figure 4.4

For a conclusion to be valid, it must always hold whenever the premises are true, i.e. for *all* possible ways of drawing the Venn diagram. Care must therefore be taken when using Venn diagrams to ensure that the most general type of diagram has been drawn.

Another important form of logical argument is known as the *law of detachment* (or sometimes as *modus ponens*). This states that if x is true and $x \rightarrow y$ is true then y is true, and may be written in the form of the schema

$$\frac{\begin{matrix} x \\ x \rightarrow y \end{matrix}}{y}$$

The proof of this argument is demonstrated in the truth table of Table 4.19 showing that $f = [x \ \& \ (x \rightarrow y)] \rightarrow y$ is a tautology.

Table 4.19. Truth table of $f = [x \ \& \ (x \rightarrow y)] \rightarrow y$

x	y	$x \rightarrow y$	$x \ \& \ (x \rightarrow y)$	f
0	0	1	0	1
0	1	1	0	1
1	0	0	0	1
1	1	1	1	1

It is instructive to consider the following three examples:

Schema

(a) If x^2 is odd, then x is odd $p \rightarrow q$

$$\frac{x^2 \text{ is odd}}{x \text{ is odd}} \qquad \frac{p}{q}$$

(b) If x is even, x^2 is even $r \rightarrow s$

$$\frac{x \text{ is odd (i.e. not even)}}{x^2 \text{ is odd}} \qquad \frac{\bar{r}}{\bar{s}}$$

(c) If 3 is even, 3^2 is even $u \rightarrow v$

$$\frac{3 \text{ is even}}{3^2 \text{ is even}} \qquad \frac{u}{v}$$

Example (a) is a straightforward application of the law of detachment yielding a correct conclusion that the square root of an odd number is itself odd. It should be mentioned that any valid argument remains valid if a specific proposition is substituted for *every* occurrence of a propositional variable (*rule of substitution*). Thus, if in (a) '3^2 is odd' is substituted for p and '3 is odd' is substituted for q, the argument becomes

$$\frac{\begin{matrix} \text{If } 3^2 \text{ is odd, then 3 is odd} \\ 3^2 \text{ is odd} \end{matrix}}{3 \text{ is odd}}$$

giving the valid conclusion '3 is odd' from two true premises.

Example (b) is not an instance of the law of detachment as is evident from its schema. It is in fact not a valid conclusion from the two premises, although it is a trap into which many people fall. In this particular case, the conclusion happens to be correct but it does not follow logically from the two true premises due to a fallacious argument.

Example (c) is again an instance of the law of detachment and is a valid logical argument although it leads to the surprising conclusion '3^2 is even'. This arises from the use of the false premise '3 is even'. From these three examples, it is seen that the validity of a logical argument does not depend in any way on the truth or falsity of the conclusion.

More complicated arguments can often be verified by using a combination of the law of syllogism and the law of detachment. The schema

$$\bar{x} \vee y$$
$$\bar{y} \vee z$$
$$\underline{x}$$
$$z$$

may be verified by first noting that

$$\bar{x} \vee y = x \to y$$

and then proceeding by the following steps:

$$x \to y \quad \text{premise } \bar{x} \vee y$$
$$\underline{y \to z \quad \text{premise } \bar{y} \vee z}$$
$$x \to z \quad \text{law of syllogism}$$
$$\underline{x \quad\quad\quad \text{premise}}$$
$$z \quad\quad\quad \text{law of detachment}$$

Suggestions for Further Reading

Carroll, Lewis, 1955, *Symbolic Logic*, MacMillan, New York.
Church, A., 1956, *Introduction to Mathematical Logic*, Princeton University Press, Princeton, N.J.
Freudenthal, H., 1966, *The Language of Logic*, Elsevier, Amsterdam.
Kneebone, G. T., 1963, *Mathematical Logic*, Van Nostrand, London.
Mendelson, E., 1964, *Introduction to Mathematical Logic*, Van Nostrand, London.
Rosenbloom, P., 1950, *The Elements of Mathematical Logic*, Dover Publications, New York.
Wilder, R. L., 1952, *Introduction to the Foundation of Mathematics*, Wiley, New York.

Exercises

1. State which of the following are propositions, and write in symbolic
   language those that are propositions using letters to denote each elementary
   proposition.
   (a) Is grass green?
   (b) Grass is yellow
   (c) Newly cut grass
   (d) Snow is white and the moon is made of cheese
   (e) If four is a prime number then eight is prime
   (f) Either I shall go out tonight or I shall stay in
   (g) You must study this lesson
   (h) Plants will grow if and only if it is warm and it is wet

2. Let
       c denote today is cold
       r denote it is raining
       s denote yesterday it was sunny
   Translate into English the following:
   (a) $c \& r$
   (b) $r \rightarrow c$
   (c) $s \rightarrow (\bar{c} \& \bar{r})$
   (d) $(c \downarrow r) \equiv s$
   (e) $\overline{c \& r \& s}$
   (f) $(c \not\equiv r) \& \bar{s}$

3. Write out the complete truth tables for the following compound statements:
   (a) $(x \& y) \equiv (x \vee y)$
   (b) $(\bar{x} \equiv \bar{y} \equiv \bar{z}) \& (x \downarrow y \downarrow z)$
   (c) $((w \rightarrow x) \rightarrow y) \rightarrow z$
   (d) $(w \vee \bar{w} \vee \bar{x} \vee x) \& (x \rightarrow y) \& x \& \bar{y}$

4. Give the special disjunctive and special conjunctive normal forms from the
   truth tables obtained in Question 3.

5. Demonstrate the theorems of Boolean algebra that were given in Chapter 3
   by writing out the appropriate truth tables.

6. Show that the operation & is not complete by itself.

7. Prove that the following are tautologies, by writing out the truth tables and
   also by drawing Venn diagrams:
   (a) $x \rightarrow x$
   (b) $\bar{x} \rightarrow (x \rightarrow y)$

(c) $(x \& y) \rightarrow x$

(d) $(x \rightarrow y) \rightarrow ((z \lor x) \rightarrow (z \lor y))$

(e) $(x \rightarrow y) \lor (y \rightarrow x)$

(f) $x \rightarrow (y \rightarrow x)$

(g) $x \& (x \rightarrow y) \rightarrow y$

8. Show by means of a Venn diagram that if $x \rightarrow y$ then $\bar{y} \rightarrow \bar{x}$. This is known as the contrapositive of the proposition. What is the contrapositive of the proposition 'All students are poor'?

Let

y be the proposition 'x^2 is odd'

z be the proposition 'x is odd'

Prove that z is a valid conclusion from y (i.e. $y \rightarrow z$) by demonstrating the contrapositive. (A conditional statement and its contrapositive are logically equivalent.)

9. The converse of $x \rightarrow y$ is $y \rightarrow x$. Find examples of propositions to show that the converse does not necessarily hold true and also an example for which the converse is true.

10. Investigate the validity of the following arguments by using Venn diagrams, and give examples to illustrate the results.

(a) All x are y
All z are not y
_____
All z are not x

(b) Some x are y
All z are y
_____
Some x are z

(c) All x are y
No w is z
All y are z
_____
No x is z

11. Determine the validity of the following arguments
(i) by using truth tables,
(ii) by deduction from the law of syllogism, law of detachment and the contrapositive.

(a) $x \rightarrow \bar{y}$
$z \rightarrow y$
z
_____
$\bar{x}$

(b) $x \rightarrow y$

$x \vee z$

$\bar{z}$

___

y

(c) $x \rightarrow y$

$z \rightarrow \bar{y}$

___

$z \rightarrow \bar{x}$

12. Express the following in terms of the algebra of propositions and prove their validity.

(a) The rules governing the choice of a sovereign of a country are

   (i)   All sovereigns are kings or queens

   (ii)   All kings are of royal blood

   (iii) No queen may become sovereign

___

   Every sovereign must have royal blood

(b) A computer specification contains these requirements for its circuits:

            (i)   They are transistorized

(AND) (ii)   They are on printed boards

            (iii) No transistor circuit is unreliable

            (iv) All printed circuits are reliable

___

            All computer circuits are reliable

Chapter 5
Switching Circuits

Simple Switches

A simple switch as shown in Figure 5.1 has the fundamental property that it is either *on* or *off*. When it is on, A and B are connected together and current can flow. This is therefore a two-state logical device. The *transmission* function of

Figure 5.1

the circuit may be denoted by 0 for an open circuit with no transmission (off) and by 1 for a closed circuit (on). The opposite convention, 1 for off and 0 for on, is sometimes employed in which case it is called a *hindrance* function. The hindrance function is the dual of the transmission function and there is thus no need to consider it separately. Henceforth, the design of switching networks will be in terms of the transmission function.

A single switch will be denoted by a letter or variable $x, y, z, \ldots$, etc., and the circuit of Figure 5.2 has the transmission function x.

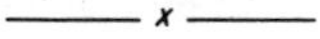

$$\text{———} x \text{———}$$

Figure 5.2

When $x = 1$, the circuit is closed; when $x = 0$, the circuit is open.

The negation of this circuit which will be closed when $x = 0$ and open when $x = 1$ is then shown as a switch $\bar{x}$ as in Figure 5.3.

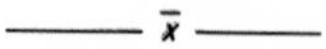

$$———\ \bar{x}\ ———$$

Figure 5.3

If two switches x, y are connected in series, the circuit of Figure 5.4 is obtained. The transmission function is shown by a truth table, Table 5.1. The circuit is closed or conducting if and only if both the switches are closed. The

Table 5.1

x	y	f
0	0	0
0	1	0
1	0	0
1	1	1

$$——— x — y ———$$

Figure 5.4 Series connection

truth table is identical with the operation of conjunction (&) in the algebra of propositions and to the Boolean product operation (.). The transmission function f of two switches in series may therefore be written as

$$f = x\ \&\ y \text{ or } x \cdot y$$

The parallel connection of two switches is shown in Figure 5.5 and the appropriate truth table in Table 5.2. This truth table is identical with disjunction

Table 5.2

x	y	f
0	0	0
0	1	1
1	0	1
1	1	1

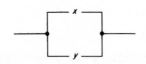

Figure 5.5 Parallel connection

(V) and to the Boolean sum operation (+). The transmission function for two switches in parallel is

$$f = x \lor y \text{ or } x + y$$

Two switching circuits are equivalent if they have the same transmission functions and this may be determined by using truth tables in the same manner as for the algebra of propositions.

It is now easy to check that the axioms of a Boolean algebra are satisfied by the transmission functions of switching circuits.

There are two binary operations:

series connection (.)
parallel connection (+)

Both operations are clearly commutative,

$$x . y = y . x$$
$$x + y = y + x$$

Identity elements 0 and 1 correspond to an open circuit and a short circuit respectively. From Figure 5.6, it can be seen that

$$x . 1 = x$$
$$x + 0 = x$$

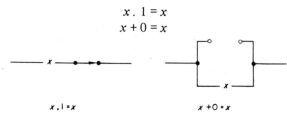

Figure 5.6

The truth of the distributive axioms may be demonstrated by evaluating the transmission function of the circuits shown in Figure 5.7.

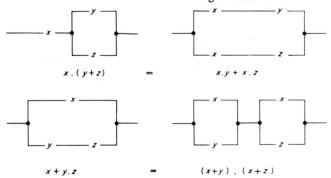

Figure 5.7

An inverse element exists. For a switch x this is the negation $\bar{x}$. From Figure 5.8 it can be seen that

$$x + \bar{x} = 1$$
$$x . \bar{x} = 0$$

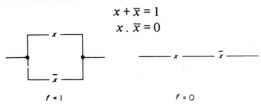

Figure 5.8 Properties of negation

Thus the algebra of circuits is a Boolean algebra. The dual system obtained by interchanging 0 and 1, and + and . , is also a Boolean algebra and, in transforming to the dual, series circuits become parallel circuits and vice versa. Thus the dual of $x + y$ is x . y (Figure 5.9). All the theorems of Boolean algebra and

Figure 5.9 Dual circuits

methods of simplification may therefore be applied to switching circuits. The truth of these theorems can usually be seen immediately from an inspection of the switching circuits. The complete set of 16 possible switching functions of two variables can be obtained by combinations of series and parallel switches. Each function can be expressed in terms of + and . , as in the algebra of propositions. From this form of the function, a switching circuit can always be drawn immediately.

The methods of Boolean algebra were first applied to switches and electric circuits by *Claude Shannon* in 1938. Since that time, the algebra of switching circuits has received a great deal of attention from both mathematicians and engineers and found widespread applications in the design of digital computers, telephone switching systems, and various electronic control systems.

Some Examples of Switching Circuits

(i) What is the transmission function of the switching circuits shown in Figure 5.10? Circuit (a) has switching elements y, $\bar{x}$ in parallel to give $\bar{x} + y$ and this is in series with switch x. Therefore

$$f = x \,.\, (\bar{x} + y)$$

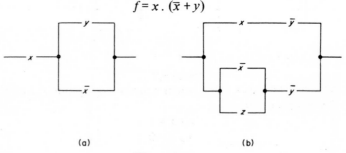

(a)   (b)

Figure 5.10

Using the distributative law,

$$f = x \,.\, \bar{x} + x \,.\, y$$
$$= x \,.\, y$$

For circuit (b), x, $\bar{y}$ in series $(x \cdot \bar{y})$ are in parallel with $(\bar{x} + z)$ in series with $\bar{y}$.

$$\begin{aligned}
f &= x \cdot \bar{y} + (\bar{x} + z) \cdot \bar{y} \\
&= x \cdot \bar{y} + \bar{x} \cdot \bar{y} + \bar{y} \cdot z \\
&= (x + \bar{x}) \cdot \bar{y} + \bar{y} \cdot z \\
&= \bar{y} \cdot (1 + z) \\
&= \bar{y}
\end{aligned}$$

(ii) Draw a switching circuit to realize the transmission function

$$f = x \cdot y + \bar{x} \cdot (\bar{y} + x + y)$$

This expression could be simplified first by Boolean algebra but, for the purpose of this example, the switching circuit for the complete expression will be drawn.

Start at the innermost bracket and form $\bar{y} + x + y$, i.e. $\bar{y}$, x, y in parallel. This must be in series with a switch $\bar{x}$ and this complete section is in parallel with x and y in series $(x \cdot y)$. The complete circuit is shown in Figure 5.11.

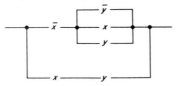

Figure 5.11

(iii) Draw a switching circuit to produce as an output the propositional function

$$f = (\bar{x} \to y) \mathbin{\&} (y \not\equiv z)$$

There are two possible approaches. Firstly replace each operation by its equivalent in terms of $\cdot$ and $+$ connectives.

$$\begin{aligned}
x \to y &= \bar{x} + y \\
x \not\equiv y &= x \cdot \bar{y} + \bar{x} \cdot y
\end{aligned}$$

Thus

$$f = (x + y) \cdot (y \cdot \bar{z} + \bar{y} \cdot z)$$

If a switching circuit with a transmission function f is produced and a logic level 1 applied at the input, then the output will represent the given Boolean function f. This is shown drawn in Figure 5.12.

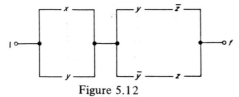

Figure 5.12

As an alternative method, the truth table may be completed and the special disjunctive normal form obtained. This gives the result

$$f = \bar{x} \cdot y \cdot \bar{z} + x \cdot \bar{y} \cdot z + x \cdot y \cdot \bar{z}$$

from Table 5.3. The circuit is shown in Figure 5.13.

Table 5.3

x	y	z	$\bar{x} \to y$	$y \neq z$	f
0	0	0	0	0	0
0	0	1	0	1	0
0	1	0	1	1	1
0	1	1	1	0	0
1	0	0	1	0	0
1	0	1	1	1	1
1	1	0	1	1	1
1	1	1	1	0	0

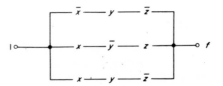

Figure 5.13 Switching circuit for $(\bar{x} \to y)$ & $(y \neq z)$

(iv) Design a circuit to control a light so that the light may be switched on and off independently by any one of three switches.

The action of the switches must be that changing the state of any one switch changes the state of the light to off if it is on, and to on if it is off. Assume that the initial state of the system is with the light off when all three switches are off. The transmission function of the circuit is given by the truth table of Table 5.4.

Table 5.4

x	y	z	f
0	0	0	0
0	0	1	1
0	1	0	1
0	1	1	0
1	0	0	1
1	0	1	0
1	1	0	0
1	1	1	1

The special disjunctive normal form is therefore

$$f = \bar{x}\,.\,\bar{y}\,.\,z + \bar{x}\,.\,y\,.\,\bar{z} + x\,.\,\bar{y}\,.\,\bar{z} + x\,.\,y\,.\,z$$

This gives the circuit shown in Figure 5.14.

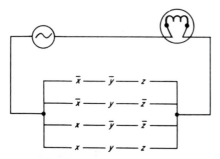

Figure 5.14 Light circuit

Simplification of Switching Circuits

Any of the theorems of Boolean algebra may be used to simplify the transmission function of a switching circuit together with the methods that will be presented in the next few chapters.

With simple series-parallel circuits it is often quite easy to simplify a circuit by inspection, by considering all possible transmission paths. As an illustration, the previous set of examples will be considered.

(i) From Figure 5.10(a), it is obvious that there cannot be a transmission path through the switches x, $\bar{x}$. Thus the only path is through the two switches x, y giving a simplified circuit consisting of these two switches in series with a transmission function $x\,.\,y$.

For the circuit of Figure 5.10(b), both of the parallel branches include a switch $\bar{y}$. As a first simplification, this can be performed with one switch $\bar{y}$

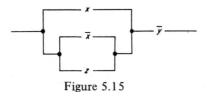

Figure 5.15

placed as in Figure 5.15. This leaves the switches x, $\bar{x}$, z in parallel which must always be a short circuit. The circuit therefore reduces to a single switch $\bar{y}$.

(ii) There are three switches $\bar{y}$, x, y in parallel which have a transmission function 1 and this gives the circuit of Figure 5.16.

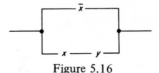

Figure 5.16

The final part of this simplification is seen more easily by using Boolean algebra:

$$f = \bar{x} + x \cdot y$$
$$= (\bar{x} + x) \cdot (\bar{x} + y)$$
$$= 1 \cdot (\bar{x} + y)$$
$$= \bar{x} + y$$

This represents two switches $\bar{x}$, y in parallel. Having determined the simplification by algebra, it is easy to verify by inspection of the transmission paths.

(iii) Figure 5.12 can be simplified by noting that the transmission is unchanged if the first y switch is omitted as in Figure 5.17.

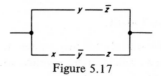

Figure 5.17

With the alternative circuit of Figure 5.13, it is possible to combine the two x switches into a single switch and also the two $\bar{z}$ switches as in Figure 5.18. This particular circuit is interesting in that it is not a series-parallel circuit. This is seen more clearly if it is redrawn as in Figure 5.19. This form of circuit is an example of a *bridge* circuit.

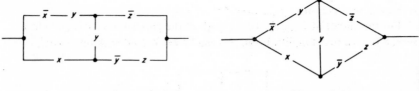

Figure 5.18   Figure 5.19

A bridge equivalent of a series-parallel circuit can often be obtained using fewer switching elements. There is, however, no general method of obtaining a bridge equivalent using fewer elements than the original series-parallel circuit. Some techniques for manipulating non-series-parallel circuits will be given later.

In this example, however, the bridge circuit does not give the minimum, since Figure 5.19 can be further simplified. There are two paths which lead through the $\bar{z}$ switch. These are $\bar{x} \cdot y$ along the top arm and $x \cdot y$ along the bottom and vertical arms. This only depends on y and not on x, and so it reduces to the same circuit as Figure 5.17.

(iv) Two switches can be saved from Figure 5.14 to give Figure 5.20.

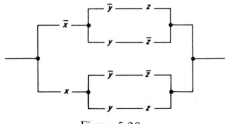

Figure 5.20

Negation of Switching Circuits

A switching circuit is frequently required for a Boolean function which contains the negation of various combinations of elements. While it is of course possible to manipulate the expression by applying De Morgan's laws and thus obtain only negations of single elements, it is often more convenient to be able to draw a negated circuit directly. The graphical method to be shown consists essentially of forming the dual network, i.e. change parallel connections to series and vice versa, and then negating each of the variables.

The method is illustrated in Figure 5.21 for the Boolean function

$$f = \overline{p \cdot (\bar{q} + r \cdot \bar{s})}$$

Thus, if the circuit for $\bar{f} = p \cdot (\bar{q} + r \cdot \bar{s})$ is drawn and then negated, the required function is obtained. There are four steps.

(a) Draw the circuit so that all the lines with switches are parallel.

(b) Draw lines through each switch at right angles to the original lines.

(c) Join the new lines to form the completed circuit. In doing this, each mesh or loop of the original circuit should contain a single node of the new circuit. The new input and output nodes lie outside the original circuit, one on each side of a line through the original input and output nodes. The connecting leads are considered to be extended indefinitely.

(d) Finally, negate each of the individual switching elements, i.e. p becomes $\bar{p}$, $\bar{q}$ becomes q, etc.

Logic Elements

The switching elements considered so far only had the property of connecting two or more points of a circuit together, thus allowing current to flow. An alternative type of logic element is one that produces an electrical output voltage when a specified combination of voltages is present at the input. For a two-value logic system which is the one in general use, there are only two possible voltage levels, denoted by 0 and 1.

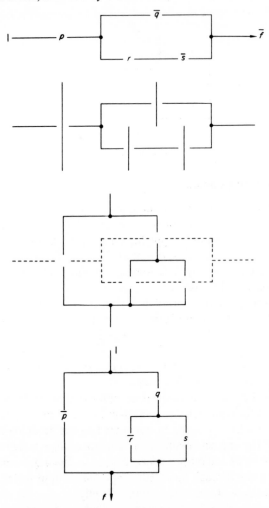

Figure 5.21 Negation of switching circuit

There are two possible systems. In the first, which is known as *positive* logic, 1 represents a voltage which is more positive than the voltage represented by 0. For example, 1 might represent zero volts and 0 represent −10 volts. In the other system, *negative* logic, 1 represents a more negative voltage than 0, e.g. 1 represents −10 volts, 0 represents zero volts. As can be seen, the actual voltages involved are immaterial.

It is possible to build circuit elements to perform any of the propositional connectives. Unlike the switches, these elements do not operate in both directions and signals pass only from input towards output. There can be 1, 2, 3, or more inputs to an element, depending on the type of connective, the maximum number allowed being referred to as the *maximum fan-in*. The elements for Boolean product (.) and sum (+) are commonly called AND gates and OR gates respectively. Since any Boolean function can be expressed entirely in terms of the pairs (¯, .) or (¯, +), circuit elements for these operations alone will enable any logic circuit to be constructed. The same is also true for NOR gates alone or NAND gates.

British Standard Symbols

In March 1969, the British Standards Institution issued a new standard for logic symbols entitled *Pure Logic and Functional Symbols* and issued as BS 3939: Section 21. This replaces an earlier standard, viz. Supplement 5 of BS 530, and has been introduced to bring the British symbols into agreement with the International Electrotechnical Commission. Since there are serious differences between the two sets of symbols, the new B.S. symbols will be described first, followed by a brief description of the differences.

Figure 5.22 B.S. gate

The basic symbol for a gate is a semi-circle as shown in Figure 5.22, with an appropriate number of inputs. If there are more input connections than can be drawn conveniently, either of the symbols of Figure 5.23 may be used.

Figure 5.23 Multi-input AND gate

Any output connection should leave the gate in a radial direction. There must be a distinguishing character inside the semi-circle to denote the type of gate. For instance, an AND gate has the symbol & inside.

A general element which may be used for several particular gates is the *threshold gate.* This switches when a given threshold level of inputs is reached; hence its name. That is to say that, when more than some specified number of input lines have a logic 1 on them, the output gives a logic 1. If fewer inputs than this number are 1s, then the output remains at 0. The required minimum number of inputs for switching to occur is written inside the symbol. Thus Figure 5.24 indicates that two or more inputs must be 1 for the output to be 1.

Figure 5.24 Threshold gate

An OR gate is a special case of the threshold gate for which the output is 1 if and only if *one* or more of the inputs are 1. It is therefore represented by a 1 inside the semi-circle (Figure 5.25). The AND gate can also be considered as a particular case of a threshold element in which the number inside the symbol must be equal to the total number of input connections.

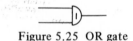

Figure 5.25 OR gate

The negation of signals is often combined with the other operations of a gate. It is represented by a small *circle* across the appropriate signal line, and may be a negated input or a negated output. Thus Figure 5.26(a) performs the operation $\bar{x} . \bar{y}$; Figure 5.26(b) gives the equiveridic output $\overline{x + y}$ or $x \downarrow y$ and is the B.S.

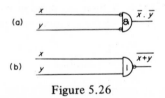

Figure 5.26

symbol for a NOR gate. Figure 5.27 is an AND gate which gives a negated output $\overline{x . y}$ in addition to the normal output $x . y$. A negater, when required as a separate element, is drawn as a NOR gate with one input as in Figure 5.28.

Figure 5.27   Figure 5.28 Negater

A gate may also have an *'inhibit'* input which has the effect of preventing the output becoming 1 whenever the inhibiting line is 1, no matter what the values of the other input signals. It is represented on the symbol by a bar across the line. Thus, for example, the x input in Figure 5.29 is an inhibit signal, and the output of the gate may be expressed as $\bar{x} \cdot (y + z)$.

Figure 5.29 Inhibit signal

The IMPLIES operation is non-symmetric since $x \rightarrow y$ and $y \rightarrow x$ represent different functions. Any symbol for this operation must therefore distinguish between the two inputs. There is no B.S. symbol specially for IMPLIES but, since $x \rightarrow y = \bar{x} + y$, the symbol shown in Figure 5.30 represents the same

Figure 5.30

operation and will be used when necessary in this book. The standard also allows special gates by inserting some character in the symbol with an explanatory note on the diagram to denote the special properties. As an example, the logic circuit for the propositional function

$$f = (\bar{x} \rightarrow y) \,\&\, (y \not\equiv z)$$

could be drawn as in Figure 5.31, where the symbol $\not\equiv$ has its usual meaning of NON-EQUIVALENCE.

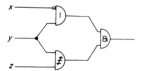

Figure 5.31

Provision is made to indicate, where necessary, whether positive or negative logic is being used. This choice may be shown by either (a) stating on the logic diagram whether positive or negative logic is being used exclusively, or (b) using an indicator to denote negative logic on any particular input or output. The presence of the triangular-shaped indicator, as shown in Figure 5.32, denotes

Figure 5.32 Negative logic output

that the 1 state is the more *negative* digital level (i.e. negative logic). If the symbol is absent on any input or output then the 1 state is the more *positive* digital level. The OR gate of Figure 5.32 thus has inputs using positive logic while the output is in negative logic.

In the earlier B.S. symbols of BS 530, logic elements were represented by circles, with radial input connections on one side of a diameter, as shown in

Figure 5.33. A NOR gate was generally shown as an example of a threshold gate with the symbol T inside the circle (Figure 5.34) denoting that the output would be high if *not one* of the inputs were high.

Figure 5.33 Old B.S. gate   Figure 5.34 Old B.S. NOR gate

The most significant differences, however, are that (a) negation was represented by a bar, and (b) inhibition was represented by a small circle. That is to say, the meanings of the two symbols representing negation and inhibition have been reversed in the new standard.

Electronic Gates

This book is concerned with the design of logic circuits and not the detailed electronic design of individual gating elements. However, a few general remarks are necessary as the type of gates available and their performance as actual physical elements can affect the design of a logic circuit.

A very simple AND gate or OR gate may be constructed using only resistors and diodes together with a voltage source as in Figure 5.35. Since these are

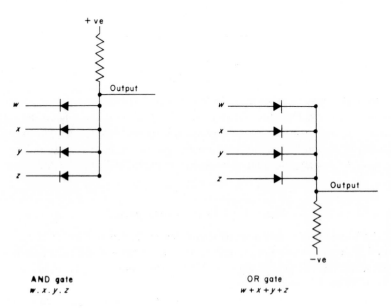

Figure 5.35 Passive network gates (positive logic)

passive devices, there is an attenuation of the input signal in passing through the gate and normally a further loading effect when the output is fed into the input of the following gate in the circuit. The speed of switching, although much faster than a relay, is rather slow by electronic standards. If a gate is to feed several loads without affecting the speed and accuracy of the overall system, it will generally be necessary to use active devices for the gating circuits. For any given electronic gate there will be a maximum number of outputs which may be taken from it without affecting its ability to keep within its specification. This is referred to as the *maximum fan-out* of the element.

Transistor AND gates and OR gates can be made by replacing the diodes in the previous circuit of Figure 5.35 with transistors. This type of circuit thus requires one transistor for every input to a gate. A much more economical circuit may be obtained if NOR gates (or their dual-circuit NAND gates) are formed. A NOR gate can be constructed with a diode for each input signal, essentially giving the Boolean sum of the inputs, together with a single transistor which both acts as an inverter and provides the necessary gain. The diodes also prevent the transistor saturating and the overall circuit provides a cheap, fast, electronic gate. This type of circuit is known as diode—transistor logic (DTL). All the propositional connectives can be obtained using NOR gates only (or NAND gates only), and since these can now be manufactured very cheaply they have become widely available and are very popular with logic designers. The design of logic circuits using NOR circuits will be discussed in the next chapter.

The logic gates described so far have been constructed from discrete components such as diodes, resistors, and transistors. These may often be obtained mounted on printed circuit boards and then encapsulated in epoxy resin. However, another form of construction known as an *integrated circuit* is now very popular. In an integrated circuit, all the components required for a specific logic gate are formed directly on layers of semiconductor material by diffusion or growth process.

Owing to mass production, integrated circuits can be produced very economically and also have the advantages of small size and high operating speeds. The performance of DTL circuits discussed above can in fact be improved still further by effectively replacing the input diodes with transistors. The circuits are then known as transistor—transistor logic (TTL) and possess faster switching speeds, high noise immunity, and good drive characteristics. In addition many multiple type gates can be produced, for example the AND/NOR gate which gives an output $w\,x \downarrow y\,z$.

Whatever type of gate is used, however, there will always be a slight delay of a signal in passing through a gate. In evaluating the performance of a logic circuit, it is often necessary therefore to know how many gates a signal has to pass through. This is known as the *logical level*. Since the delay is in general proportional to the logical level, a circuit with a low logical level is preferable to one with a high level. In this connection NEGATERS are not always counted as separate logical levels as this operation is frequently possible at the input or output of other gates.

References and Further Reading

Chu, Y., 1962, *Digital Computer Design Fundamentals*, McGraw-Hill, New York.
Hohn, F. E., 1966, *Applied Boolean Algebra*, 2nd edn., MacMillan, New York.
Shannon, C., 1938, 'A symbolic analysis of relay and switching circuits', *Trans. AIEE*, **57**, 713–23.
British Standards Institution, BS 3939: Section 21, 1969, *Pure Logic and Functional Symbols*.
Digital Equipment Corporation, *Logic Handbook*.

Exercises

1. Obtain the complete set of 16 possible switching functions by combinations of series and parallel switches.

2. Demonstrate the following theorems of Boolean algebra by comparing the switching circuits representing each side of the identities.
   (a) $x + (y + z) = (x + y) + z$
   (b) $x + x = x$
   (c) $0 + x = x$
   (d) $0 . x = 0$
   (e) $x . (\bar{x} + y) = x . y$
   (f) $x + (x . y) = x$
   (g) $(x + y) . (\bar{x} + y) = y$
   (h) $(x + y) . (\bar{x} + z) = x . z + \bar{x} . y$

3. Draw the dual circuits for each equation of Question 2.

4. Construct switching circuits, without any previous simplification, to produce the following switching functions:

   (a) $x \cdot y + x \cdot \bar{y} + \bar{x} \cdot \bar{y}$

   (b) $w \cdot [x \cdot (y + z) + y \cdot (x + z)]$

   (c) $(w \cdot x + \bar{y} \cdot \bar{z}) \cdot (w \cdot x + y + z)$

5. Simplify directly the circuits obtained in Question 4. Prove the results are correct by Boolean algebra.

6. Use the method of graphical negation to illustrate De Morgan's laws.

7. Draw switching circuits, using graphical negation where appropriate, for the following functions:

   (a) $\overline{x \cdot y \cdot z} + \overline{\bar{x} \cdot \bar{y} \cdot \bar{z}}$

   (b) $\overline{\overline{x + y} \cdot \overline{y + z} \cdot \overline{z + x}}$

   (c) $\overline{(x \cdot y + \bar{y} \cdot z) \cdot x \cdot (y \cdot z + x \cdot \bar{z}) + x \cdot y}$

   Write out the transmission functions of the final circuits and check their correctness by means of truth tables.

8. Draw logic circuits, using British Standard symbols, for each of the expressions of Question 4 and Question 7.

9. Draw logic circuits for each of the 16 possible switching functions of two variables.

   (a) Using AND and OR elements only. Negations may be used on inputs or outputs of the gates.

   (b) Using NOT-IMPLIES elements only.

10. Using AND, OR, and NOT elements only, draw logic circuits for the following expressions:

    (a) $(x \to y) \cdot (y \to z)$

    (b) $w \equiv x \equiv y \equiv z$

    (c) $(w \downarrow x \downarrow y) \downarrow (p \downarrow q \downarrow r)$

    (d) $[(\bar{w} \to x) | (y \not\equiv \bar{z})] + (w \equiv \bar{x}) \cdot (\bar{y} \leftarrow z)$

11. Design a logic circuit to give an output f from four input variables w, x, y, z to satisfy the following conditions:

    (i) $f = 1$ when any three or more of the input variables are 1

    (ii) $f = 1$ if x and y are both 1 and z is 0

    (iii) $f = 1$ if $w \equiv z = 1$

    (iv) $f = 1$ if $x \to y = 1$

    (v) $f = 0$ otherwise

12. Design a logic circuit to represent the following problem. An output should be given from the circuit to drive a warning light when any 'unsafe' combination occurs.

A man is taking a goat, a fox, and a cabbage to market when he comes to a river. The only boat available will carry only one object apart from himself. The goat should not be left alone with the fox nor may the goat be left with the cabbage.

Chapter 6
NOR / NAND Gates

Introduction

NOR/NAND gates have become very popular elements for use in the construction of logic circuits. As discussed in the previous chapter, they can be manufactured very cheaply and therefore have become widely available. Since all logic functions can be expressed entirely in terms of either the NOR or the NAND connective, logic designers can produce circuits using just one type of gate.

There is no essential difference between the two units. NOR and NAND are dual operations. In an actual circuit element, if a gate performs the NAND operation with positive logic (i.e. with the higher level of potential denoted by 1 and the lower as 0), then with negative logic (0 higher level than 1) the NOR operation is obtained. To avoid undue repetition, only the design of circuits using NOR units will be discussed in detail. The corresponding circuits for NAND gates can be obtained by duality if required.

A NOR gate with no inputs will always have a logic 1 as its output. If a single input x is applied, then the output is

$$x \downarrow 0 = \overline{x + 0}$$
$$= \overline{x}$$

Thus a NOR gate with a single input acts as a negater (Figure 6.1).

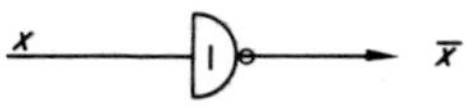

Figure 6.1

As mentioned earlier, the NOR operation is not associative. The output from the circuit shown in Figure 6.2, with three inputs is

$$x \downarrow y \downarrow z = \overline{x + y + z}$$

If $(x \downarrow y) \downarrow z$ is required, the circuit of Figure 6.3 should be used.

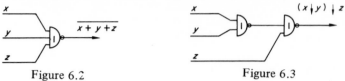

Figure 6.2   Figure 6.3

Any unused inputs of a multi-input NOR gate must have logic 0 on them. In a similar manner, when using NAND gates, unused inputs must have logic 1 applied. This is frequently included as part of the gate design so that any unconnected input automatically has the appropriate logic signal.

Design of NOR Circuits

The algebraic laws of the NOR operation are not very convenient and make the manipulation of expressions in terms of $\downarrow$ rather cumbersome. Most of the design work is generally done in terms of . and +, which are easily manipulated, and then converted to a NOR circuit. There are several possible techniques for this conversion. If the truth tables are developed for the two circuits of Figure 6.4, they both give the same set of truth values. Thus a circuit can be obtained for the two-level circuit of Figure 6.4(a) using an identical configuration of NOR units. With circuits drawn from normal forms, it is possible to obtain transformation techniques based on these equalities to give the NOR representation. This method is not very flexible, however, and more direct techniques will be employed.

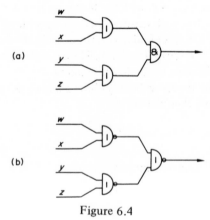

Figure 6.4

An expression in terms of the NOR operator is easily obtained using Boolean algebra if the required function is expressed initially in either a conjunctive or a disjunctive normal form. The usual starting point will be to obtain the simplest disjunctive normal form or conjunctive normal form. This can be done by Boolean algebra but easier methods will be shown later.

The minimum circuit configuration is not easy to define since the total number of gates used depends on the permitted fan-in and fan-out, i.e. the maximum number of inputs and outputs of the available gates. It will also depend on whether variables are available in negated form or if an extra gate is required for each negation. Since in conventional AND/OR circuits negations are usually available, it has become customary not to count negation as a logic level. This distinction is more difficult to observe in the case of NAND/NOR circuits as a separate gate is required for a negation with a consequent propagation delay. Therefore, in this section, a theoretical logic level will be used including negation and with no restrictions on fan-in and fan-out. In the section on p. 112, entitled 'Church's brackets', restrictions on the fan-in will be considered. If in any practical application the count of negations is not required, it is very easily omitted.

NOR Circuits from Normal Forms

To produce a NOR circuit starting from a conjunctive normal form will require at most a theoretical logic level of three gates. The disjunctive normal form will generally require one extra level.

To convert a conjunctive normal form to NOR operations, De Morgan's laws may be used as follows:

$$(w + x + y \ldots) . (\overline{w} + x + y \ldots) . \ldots$$
$$= \overline{(w + x + y \ldots)} + \overline{(w + x + y \ldots)} + \ldots$$
$$= (w \downarrow x \downarrow y \ldots) \downarrow (\overline{w} \downarrow x \downarrow y) \downarrow \ldots$$

This requires one logic level to form any negations, the second level to give the bracketed terms, and the third level to connect the brackets.

If the disjunctive normal form is used as the starting point, a final negation is always required.

As an example consider the equivalence connective $x \equiv y$, for which the two normal forms are

$$(\overline{x} + y) . (x + \overline{y})$$

and

$$(x . y) + (\overline{x} . \overline{y})$$

Considering the conjunctive form first and converting it to NORs gives

$$(\overline{x} + y) \cdot (x + \overline{y}) = \overline{\overline{(\overline{x} + y)} + \overline{(x + \overline{y})}}$$
$$= (\overline{x} \downarrow y) \downarrow (x \downarrow \overline{y})$$

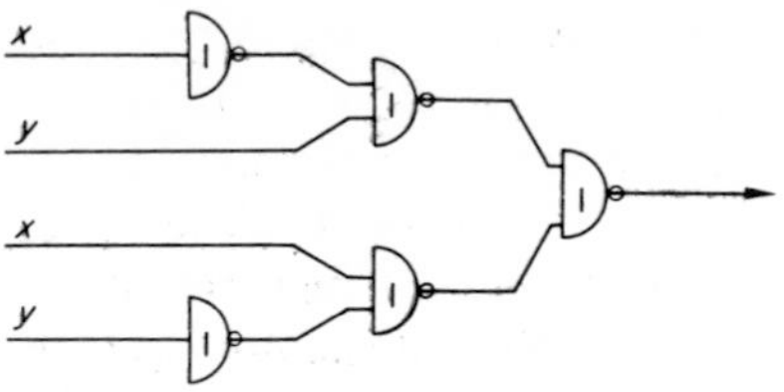

Figure 6.5 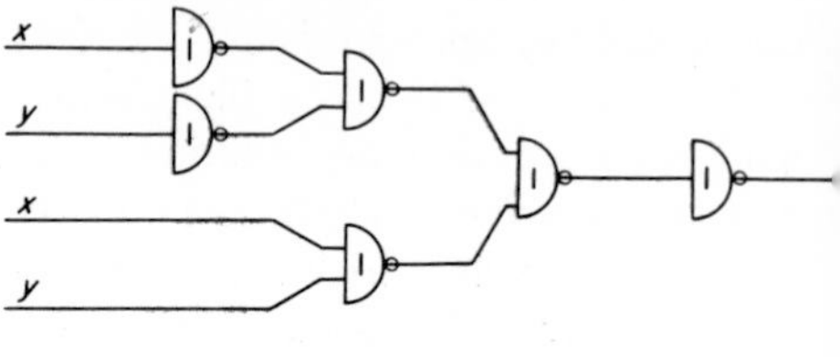   Figure 6.6

The NOR circuit is shown in Figure 6.5. The disjunctive form leads to Figure 6.6 from the equation

$$x \cdot y + \overline{x} \cdot \overline{y} = \overline{\overline{\overline{(x + \overline{y})} + \overline{(x + y)}}}$$
$$= \overline{(x \downarrow \overline{y})} \downarrow \overline{(x \downarrow y)}$$

Elimination of Redundant NOR Gates

An alternative method of obtaining a NOR diagram is to draw the circuit in terms of $^-$, ., + and then convert each of these gates into $\downarrow$ gates by using the circuits shown in Figure 6.7 which are obtained from the relations

$$\overline{x} = x \downarrow 0$$
$$x \cdot y = \overline{\overline{x} + \overline{y}} = \overline{x} \downarrow \overline{y}$$
$$x + y = \overline{\overline{x + y}} = \overline{x \downarrow y}$$

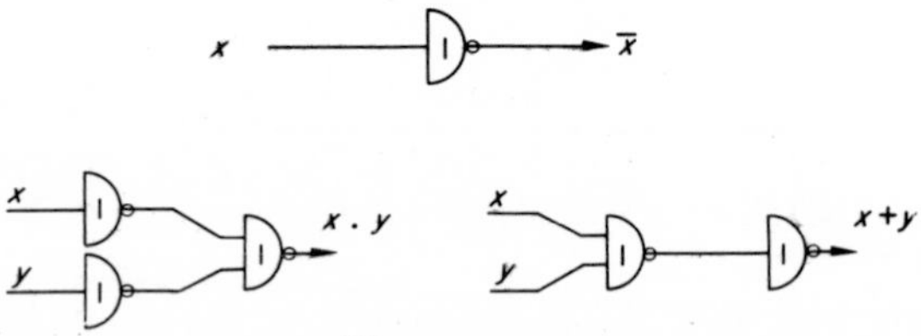

Figure 6.7

It would appear at first sight that this method would produce a circuit using two to three times as many gates as the original. However, in most instances it will be found that many of the NOR units are in fact redundant and can be eliminated. There are three rules used in simplifying NOR circuits and they are illustrated in Figure 6.8.

As an example, consider again the expression

$$x \equiv y = (\overline{x} + y) \cdot (x + \overline{y})$$

which gives the steps of Figure 6.9. It is seen that the final circuit is identical with Figure 6.5, obtained from the conversion of the conjunctive normal form.

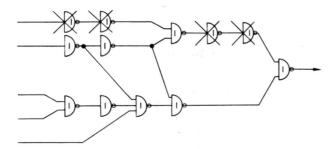

(I)   2 single-input NOR gates following one another are redundant

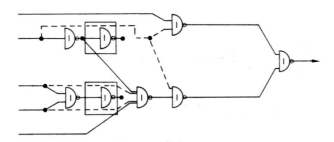

(II)   A single-input NOR gate that both feeds and is fed by NOR gates is redundant

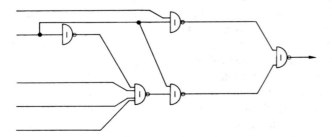

(III)   Any remaining NOR gate whose output is not now used is redundant

Figure 6.8 Reduction of NOR circuits

Since all the propositional connectives can be expressed in conjunctive normal forms they can all be converted from these into NOR forms. Thus NOR circuit equivalents of each propositional connective may be constructed. To produce a NOR circuit diagram for any Boolean function, draw first a conventional Boolean circuit using the given connectives. Convert each

individual operator into its NOR equivalent and finally simplify the resulting
diagram by removing any redundant NOR elements. This procedure is followed

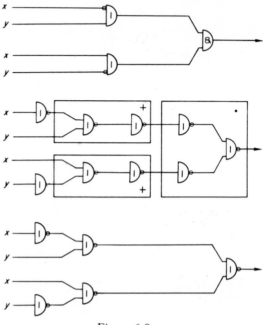

Figure 6.9

in Figure 6.10 to draw a circuit for the function $f = (\bar{x} \rightarrow y) \cdot (y \not\equiv z)$. In this
instance, the circuit obtained is identical with that drawn by algebraic conver-
sion of the *simplest* conjunctive normal form.

Circuits obtained from disjunctive or conjunctive normal forms do not
always yield the minimum number of NOR gates, even starting from the
simplest normal form. Some algebraic manipulation will frequently provide a
better circuit. Two useful points to consider are described below.

The method of transformation from AND, OR gates to NOR gates involves
the addition of extra negaters. For example, to convert AND (.) requires the
negation of each input and to convert OR (+) requires the negation of the output as was shown in Figure 6.7. Elimination of NOR gates will take place when
two such negaters follow one another as in the conversion of conjunctive
normal forms where + is followed by . . Gates will also be eliminated, however,
if the product connective is operating on negated variables as in an expression
$\bar{x} \cdot \bar{y} \cdot \bar{z}$, which only requires one NOR gate. Therefore, if the expression for
which a NOR circuit is required can be manipulated algebraically so that the
product terms consist of the conjunction of negated variables, this simplification
will then be possible.

Consider as an example the equivalence operation expressed initially in disjunctive normal form

$$x \equiv y = x \,.\, y + \overline{x} \,.\, \overline{y}$$

The final term is of the desired form but the other term $x \,.\, y$ is not. Use the

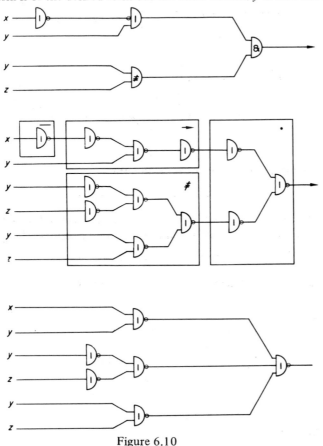

Figure 6.10

distributive law to obtain

$$(x + \overline{x} \,.\, \overline{y}) \,.\, (y + \overline{x} \,.\, \overline{y})$$

so that the product operation is acting firstly on the negated variables $\overline{x}$, $\overline{y}$ and secondly after the $+$ operation. Both instances enable inverters to be eliminated as shown in Figure 6.11. The final circuit requires only four gates as opposed to the five gates that were used in Figure 6.5 and six gates in Figure 6.6.

This example also illustrates the second method of simplification. The term $\overline{x} \,.\, \overline{y}$ is common to both factors and hence there is no need to repeat the circuit.

Whenever possible, any common terms or factors should be utilized in this manner. It is sometimes possible to obtain a common expression by inserting

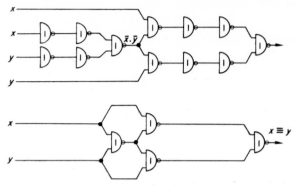

Figure 6.11

redundant terms. For example, given the function

$$f = (\bar{x} \cdot \bar{y} + z) \cdot (\bar{y} \cdot \bar{z} + x)$$

and using the equation $y \cdot \bar{z} + z = y + z$, the redundant factor $\bar{z}$ can be inserted in the first term $\bar{x} \cdot \bar{y}$ and the redundant factor $\bar{x}$ in the second. This gives

$$f = (\underline{\bar{x} \cdot \bar{y} \cdot \bar{z}} + z) \cdot (\underline{\bar{x} \cdot \bar{y} \cdot \bar{z}} + x)$$

which yields, after simplification, the circuit of Figure 6.12.

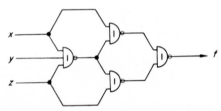

Figure 6.12

Church's Brackets

In some cases it may happen that the required number of inputs to a gate is greater than the maximum fan-in permitted by the physical construction of the unit. This will frequently happen if the gates have only two inputs. With AND or OR gates it is easy to overcome the difficulty since these operations are associative. NOR/NAND operations are not associative so that the expression

$$w \downarrow x \downarrow y \downarrow z$$

cannot be obtained by taking

$$(w \downarrow x) \downarrow (y \downarrow z)$$

It is therefore necessary to bracket the expression before converting to the $\downarrow$ operator. The expression for which the circuit is required should be fully parenthesized, each pair of brackets containing only the permitted number of variables. Suppose, for example, that only two inputs are permitted. Then, given the expression

$$x \cdot y \cdot z + \overline{x} \cdot y$$

additional brackets are then inserted, conventionally from the left-hand side of the expression, so that not more than two terms are operated upon by each connective, thus obtaining

$$(((x \cdot y) \cdot z) + (\overline{x} \cdot y))$$

The brackets inserted in this manner to parenthesize logic expressions are known as *Church's brackets* after the American logician A. Church (see Church, 1944).

A simple method of checking that equal numbers of opening and closing parentheses are present and also to find the logic levels of each operation is to count the brackets from left to right. Count the first opening or left parenthesis as 1 and increase the count by 1 each time a left parenthesis is encountered. Each time a right or closing parenthesis is encountered decrease the count by 1. Note that, if negation is to count as a logic level, an extra pair of parentheses should be placed around each negated variable. Applied to the present example, this gives the parenthesized expression

$$(\ (\ (x \cdot y) \cdot z) + (\ (\overline{x}) \cdot y) \ )$$
$$1 \ 2 \ 3 \quad\ \ 2 \quad 1 \quad 2 \ 3 \ 2 \quad 1 \ 0$$

The right-hand bracket of each pair has a count one less than its matching left-hand bracket, and the final right parenthesis should have a count of 0. The logic levels are then given by the count attached to the respective opening bracket. This concept is important when considering the time delay of the propagation of signals through gates as the highest level of operation must be completed before the next level has its correct input signals.

NOR Gates with Two Inputs Only

If the above expression $x \cdot y \cdot z + \overline{x} \cdot y$ is to be obtained using only two-input NOR gates, insert Church's brackets, *not including negation as a separate level.* This gives

$$(((x \cdot y) \cdot z) + (\overline{x} \cdot y))$$

Convert this to the ↓ operator in the normal fashion but, taking the positions of parentheses into account,

$$(((\overline{\bar{x} \downarrow \bar{y}}) \downarrow \bar{z}) \downarrow (x \downarrow \bar{y}))$$

To obtain the logic levels, the additional parentheses for the negations may now be inserted if desired

$$(\overline{ ((((\overline{(\bar{x}) \downarrow (\bar{y})})) \downarrow (\bar{z})) \downarrow (x \downarrow (\bar{y}))))$$
$$1\ 2\ 3\ 4\ 5\ 6\ 5 \quad 6\ 5\ 4\ 3 \quad 4\ 3\ 2 \quad 3 \quad 4\ 3\ 2\ 1\ 0$$

Thus the highest level is 6, when negations are counted. The circuit is shown in Figure 6.13. The usual simplifications of removing interior NOR elements must not of course be made since this would lead to more than two inputs on the following NOR unit. The only simplification that can be made here is that, since y is negated twice, one of them can be removed.

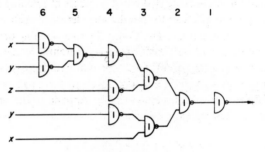

Figure 6.13

If the expression is already given in NOR notation, it is necessary to go back a stage before inserting the brackets. Thus to draw a logic circuit using two-input NOR gates for the function

$$w \downarrow x \downarrow y \downarrow z$$

rewrite it in terms of $^{-}$ and $+$ to give

$$\overline{w + x + y + z}$$

Insert Church's brackets

$$(\overline{((w + x) + y) + z})$$

and then reconvert to NORs

$$(\overline{ (\overline{(w \downarrow x)} \downarrow y)} \downarrow z)$$
$$=((((\overline{(w \downarrow x)}) \downarrow y)) \downarrow z)$$
$$1\ 2\ 3\ 4\ 5 \qquad 4\ 3 \qquad 2\ 1 \qquad 0$$

The circuit is shown in Figure 6.14.

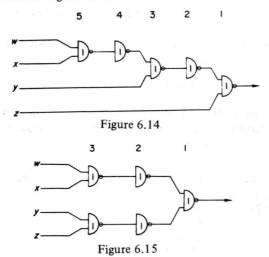

Figure 6.14

Figure 6.15

An alternative circuit which reduces the number of logic levels can be produced by inserting Church's brackets in different positions. The total number of gates is, however, unaltered in this example.

$$w \downarrow x \downarrow y \downarrow z$$
$$= \overline{w + x + y + z}$$
$$= \overline{((w + x) + (y + z))}$$
$$= \overline{((w \downarrow x) \downarrow \overline{(y \downarrow z)})}$$
$$= (((\overline{(w \downarrow x)}) \downarrow (\overline{(y \downarrow z)})))$$
$$ 1\,2\,3 \quad\;\; 2\,1 \quad 2\,3 \quad\; 2\,1\,0$$

which has a logic level of 3 and is illustrated in Figure 6.15.

References and Further Reading

Church, A., 1944, 'Introduction to mathematical logic', *Ann. Math. Studies,* **13,** Part I.

Givone, D. D., 1970, *Introduction to Switching Circuit Theory,* McGraw-Hill, New York.

Maley, G. A., and Earle, J., 1963, *The Logic Design of Transistor Digital Computers,* Prentice-Hall, Englewood Cliffs, N.J.

Exercises

1. Demonstrate the equivalence of the 2 two-level circuits shown in Figure 6.4.

2. Using the equivalence of the circuits proved in Question 1 to draw a NOR circuit for the expression

$$(p + q + r) \cdot (u + v + w) \cdot (x + y + z)$$

3. Prove algebraically the second of the reduction rules for NOR circuits shown in Figure 6.8, i.e. a single input NOR element that both feeds and is fed by NOR units is redundant.

4. Draw logic circuits for the following functions. Convert each element to NOR gates and remove any redundant elements.
   (a) $(\bar{x} + \bar{y}) \cdot (\bar{y} + \bar{z}) + \bar{x}$
   (b) $[\bar{x} \cdot \bar{y} + \bar{y} \cdot \bar{z}] \cdot \bar{x}$
   (c) $\overline{(p + q + r) \cdot (x + y + z)}$

5. Repeat Question 4 using NAND units only.

6. Draw circuits using NOR gates only, with redundant gates removed, from the AND, OR, NOT circuits that were obtained from Question 10 of Chapter 5 for the following expressions:
   (a) $(x \to y) \cdot (y \to z)$
   (b) $w \equiv x \equiv y \equiv z$
   (c) $(w \downarrow x \downarrow y) \downarrow (p \downarrow q \downarrow r)$
   (d) $(\bar{w} \to x) | (y \not\equiv \bar{z}) + (w \equiv \bar{x}) \cdot (\bar{y} \not\equiv z)$

7. Draw logic circuits for each of the 16 possible functions of two variables x, y
   (a) using NOR gates only
   (b) using NAND gates only

8. Obtain NAND circuits for each of the expressions of Question 6 by direct replacement of each operator followed by removal of redundant elements.

9. Write each of the following expressions in terms of the NOR operator ($\downarrow$) only (including the removal of negations):
   (a) $(\bar{x} + y) \cdot (\bar{y} + z) \cdot (x + \bar{z})$
   (b) $p \cdot q \cdot r + \bar{p} \cdot \bar{q} + \bar{r}$
   (c) $(x | y | z) + (\bar{x} | \bar{y} | \bar{z})$
   (d) $(p \to q) \not\equiv (x \to y)$

10. Draw circuits using the fewest number of NOR gates that you can find, making use of redundancy, etc.
    (a) $x.y.z + \bar{x}.\bar{y}.\bar{z}$
    (b) $(w.\bar{y} + \bar{x} + z).(x.\bar{z} + \bar{w} + y)$
    (c) $(y + \bar{z}).(\bar{x} + z).(\bar{y} + z)$

11. Fully parenthesize the following expressions inserting Church's brackets from the left so that no operator has more than two operands. Obtain a count of the logic levels, including negation as a separate level.
    (a) $\bar{w}.\bar{x}.\bar{y}.\bar{z}$
    (b) $w.x.y + \bar{w}.x.y + w.\bar{x}.y$
    (c) $w + x + y . \bar{w} + x + y . w + \bar{x} + y$
    (d) $\overline{w + x}.\overline{\bar{y} + z}.\overline{w + z}$

12. Draw NOR circuits with not more than two inputs to each gate for the following expressions:
    (a) $\bar{w}.\bar{x}.\bar{y}.\bar{z}$
    (b) $x \equiv y \equiv z$
    (c) $x|y|z$
    (d) $[(\bar{x} \downarrow \bar{y} \downarrow \bar{z}) \downarrow (x \downarrow y \downarrow z)]|x|y$

Chapter 7
Polish Notation and Tree Structures

Introduction

At the end of the preceding chapter, the idea of fully parenthesized expressions obtained by use of Church's brackets was introduced. Thus, with the normal precedence of operators and the conventional evaluation of expressions from the left-hand side, the propositional expression $w \lor x \lor \bar{y} \& z$ would be written as $((w \lor x) \lor ((\bar{y}) \& z))$. To the average person, this seems more difficult to understand than the original. However, if the evaluation is to be performed by a digital computer, or other automatic machine, then the fully parenthesized expression would be more convenient as it defines precisely the order of computation. There exists another method of writing algebraic expressions which preserves the syntax of the expressions but does not require any brackets. This is known as *Polish* notation from the nationality of its inventor Lukasiewicz (1963).

Lukasiewicz showed that, if a binary connective, such as &, is written in front of its two operands in the form $\& \, x \, y$ in place of $x \& y$, then by consistent use of such prefix notation no parentheses are necessary. He also introduced different symbols for the operators and the above expression $x \& y$ would have been written as $K \, x \, y$. A list of the more common connectives with their Polish equivalents is given in Table 7.1. This symbolism does have an advantage in that it can be typed on an ordinary typewriter but there is nothing to stop all conventional symbols being used with the Polish prefix notation. Indeed it can also be used with ordinary mathematical operators such as + and x (i.e. 'plus' and 'times') with $a + b$ written as $+ab$ and $a \times b$ as $\times ab$. Since the standard propositional symbols will now be familiar to the reader, they will be retained rather than the Polish alphabetic equivalents. To avoid confusion with the

Table 7.1. Polish equivalents of Boolean connectives

	Polish equivalent
$\bar{x}$	N x
$x \& y$ or $x \cdot y$	K $x y$
$x \vee y$ or $x + y$	A $x y$
$x \equiv y$	E $x y$
$x \rightarrow y$	C $x y$
$x \mid y$ or $\overline{x \& y}$	D $x y$
$x \downarrow y$ or $\overline{x \vee y}$	S $x y$
$x \not\equiv y$	R $x y$
$\overline{x \rightarrow y}$	H $x y$

normal algebraic operators, however, the connectives from the algebra of propositions (&, V, etc.) will be used throughout this chapter for logic expressions rather than the Boolean algebra symbols (. +). Since a separate symbol is required for singulary connectives, it will be convenient to write Nx for the negation of x, i.e. $\bar{x}$.

Forward Polish Notation

The expression $(x \vee y) \& z$ translates into $\& \vee x y z$ and is unambiguous since it denotes that & operates on the result of V $x y$ (i.e. $x \vee y$) and z. The unbracketed expression $x \vee y \& z$ would be represented by V $x \& y z$. Evaluation of such an expression is from the right-hand end. Move to the left until an operator is encountered. This acts on the appropriate variables to its right to give effectively a new operand. This procedure is then repeated until the complete expression is evaluated. These parenthesis-free expressions are more difficult to read than the conventional notation but are very convenient for computational purposes since no connective is encountered before both its operands are known.

As an illustration, the evaluation of

$$V \, V \, w \, x \, \& \, N \, y \, z$$

would proceed as follows. N is the first operator encountered and operates on the single variable y to give $\bar{y}$.

Thus

$$\lor \lor w\, x\, \&\, (\bar{y})\, z$$

is obtained; the next operator is $\&$ with the two operands $\bar{y}$, z giving

$$\lor \lor w\, x\, ((\bar{y})\, \&\, z)$$

Continuing in this manner gives

$$\lor\, (w \lor x)\, ((\bar{y})\, \&\, z)$$

$$(w \lor x) \lor ((\bar{y})\, \&\, z)$$

or

$$w \lor x \lor \bar{y}\, \&\, z$$

on removing redundant parentheses.

In a forward Polish notation, the order of occurrence of the variables in the expression should be the same as the order in the original. Thus, although the expressions $\&\, \lor\, x\, y\, z$ and $\&\, \lor\, y\, x\, z$ are equivalent, the first is the translation of $(x \lor y)\, \&\, z$ while the second is the translation of $(y \lor x)\, \&\, z$. However, the Polish equivalent of an expression is not necessarily unique. $x\, \&\, y\, \&\, z$ can mean either of the equiveridic expressions $(x\, \&\, y)\, \&\, z$ or $x\, \&\, (y\, \&\, z)$. These translate into Polish as $\&\, \&\, x\, y\, z$ and $\&\, x\, \&\, y\, z$ respectively.

In general, there are several possible forward Polish equivalents, but two in particular are considered. These are early-operator form in which the operators occur as early in the expression as possible and late-operator form in which they occur as late as possible. The two forms correspond to inserting Church's brackets from the left-hand side of the expression or from the right-hand side respectively. To form the early-operator form of $w\, \&\, x\, \&\, y\, \&\, z$, insert Church's brackets from the left $(((w\, \&\, x)\, \&\, y)\, \&\, z)$, giving $\&\, \&\, \&\, w\, x\, y\, z$. For the late-operator form, insert the brackets from the right $(w\, \&\, (x\, \&\, (y\, \&\, z)))$, giving $\&\, w\, \&\, x\, \&\, y\, z$. Conventional mathematical evaluation of equal precedence operators is from the left and therefore corresponds to the early-operator Polish form.

Tree Structure

Any legitimate Boolean expression may be represented by a diagram known as a tree. There exists a unique tree for any Polish expression and hence there may be several trees corresponding to any standard Boolean expression depending essentially on the position in which Church's brackets are placed. An example of the tree corresponding to the expression $\lor \lor w\, x\, \&\, N\, y\, z$ is shown in Figure 7.1. A tree consists of a set of lines called *branches* which meet at points called *nodes* or *vertices*. All the branches must be connected and there must be no closed loops or cycles. The main connective of the expression, which would provide the output of a logic circuit, is shown at the base of the diagram and is known as the *root*. The terminating vertices represent the variables and are sometimes referred to as *leaves*. Internal nodes represent connectives and each binary operator has precisely two branches emanating from it. The unary operator N has only a single branch, however. The root of the tree is at logic level 1, the next highest nodes are level 2, etc.

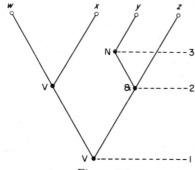

Figure 7.1

The construction of the tree from the forward Polish form is very simple. The first operator, which must be the first element in the expression, is the principal connective and is drawn as the root of the tree. Draw the correct number of branches from the root, i.e. two for a binary operator or one for a unary operator. Proceed along the left-hand branch. If the next character in the Polish string is another operator, form a new node to represent this operator and draw in the new branches. If the character in the string is a variable, then the node is a terminal and is labelled with the variable. In the present example, the first three steps give Figure 7.2. After a terminal node is

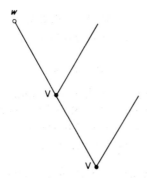

Figure 7.2

reached, it is necessary to back-track along the branches and then proceed along the first permissible (i.e. not previously used) branch encountered, which must of course be a right-hand branch. The same procedure is then carried out, continuing along new left-hand branches whenever possible and back-tracking if a terminal node is reached. The remaining steps of this example are shown in Figure 7.3, and the final tree in Figure 7.1.

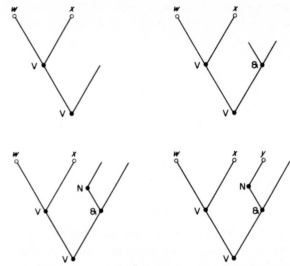

Figure 7.3

The conventional Boolean expression may be obtained immediately from the tree by commencing with the terminal nodes at the highest logic level and working down the tree so that every operand is obtained by the time it is required. The steps are as follows:

$$\bar{y}$$
$$(w \vee x) \quad ((\bar{y}) \, \& \, z)$$
$$((w \vee x) \vee ((\bar{y}) \, \& \, z))$$

or

$$w \vee x \vee \bar{y} \, \& \, z$$

This procedure in reverse enables the tree to be drawn from the Boolean expression and then from the tree the Polish form can be obtained.

The late-operator Polish form of the expression

$$\overline{w \, \& \, x \, \& \, y} \vee y \overline{\equiv z} \vee \overline{x}$$

may be found by inserting Church's brackets from the right to give, with logic levels shown,

$$(\, (\, \overline{(w \, \& \, (x \, \& \, y)}) \,) \,) \vee (\, (\, \overline{(y \equiv z)} \,) \vee (\bar{x}) \,) \,)$$
$$1 \, 2 \, 3 \qquad 4 \qquad 3 \, 2 \, 1 \quad 2 \, 3 \, 4 \qquad 3 \, 2 \quad 3 \, 2 \, 1 \, 0$$

To draw the tree, keep to the appropriate logic levels and ensure that the variables appear in the same order as in the expression. This gives the tree of Figure 7.4. Finally to obtain the corresponding Polish expression, the tree is read from the root via left-hand branches giving

$$V \ N \ \& \ w \ \& \ x y \ V \ N \equiv y z \ V \ N x$$

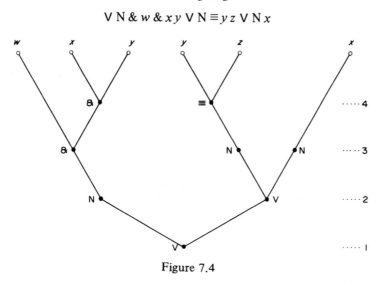

Figure 7.4

Reverse Polish Notation

To evaluate a forward Polish expression it is necessary to work from the far right-hand end of the expression. It is often much more convenient if an expression can be evaluated from the beginning or left-hand end. Of course the forward Polish string could just be written down in reverse order, e.g. the expression $x \ \& \ y \ \& \ z$ with early-operator forward Polish of $\& \ \& \ x y z$ could be written $z y x \ \& \ \&$. However, the order of the variables in the string is now in reverse order. What is actually required for the reverse Polish expression in this case is $x y z \ \& \ \&$, that is with the variables in their original order and with postfix operators. Again there are two forms of particular interest, early-operator reverse and late-operator reverse. Note in particular for non-commutative operations such as $x \to y$, forward Polish is $\to x y$ and reverse Polish is $x y \to$. The reverse Polish $y x \to$ means $y \to x$.

The tree should be drawn from the Boolean expression as before with Church's brackets from the left for early-operator and from the right for late-operator form. A Polish string is read off from the tree commencing from the root but travelling along *right*-hand branches whenever possible. The expression thus found will have the variables in the wrong order but, when this expression is reversed, the correct reverse Polish form is obtained. Thus using Figure 7.4 to

form the late-operator reverse Polish expression for $\overline{w\ \&\ x\ \&\ y} \vee \overline{y \equiv z} \vee \overline{x}$
gives first of all

$$\vee \vee N\, x\, N \equiv z\, y\, N\ \&\ \&\ y\, x\, w$$

Writing this in reverse leads to

$$w\, x\, y\ \&\ \&\ N\, y\, z \equiv N\, x\, N\, \vee \vee$$

Early-operator reverse Polish is particularly useful for computational purposes as the operators appear precisely in the required order. Consider the elementary arithmetic computation

$$2, 4, +, 3, \div, 5, 2, -, \times$$

Starting at the left gives the two operands 2, 4 followed by their operator + with the result 6; then 3 ÷, i.e. 2; continuing to the right there is 5, 2, − giving 3; and finally the preceding 2, 3, x, i.e. 6. This particular form of Polish notation is frequently used in the construction of compilers for digital computers. In fact some computers have been built with a type of store known as a 'push-down' store that effectively utilizes a language of this type.

References and Further Reading

Hamblin, C. L., 1962, 'Translation to and from Polish notation', *Comput. J.,* 5, No. 3, March, 210–13.
Lukasiewicz, J., 1957, *Aristotle's Syllogistic,* 2nd edn., Oxford University Press, Oxford.
—— 1963, *Elements of Mathematical Logic* (English translation from Polish), Pergamon Press, Oxford.
Prior, A. N., 1955, *Formal Logic.* Oxford University Press, Oxford.
Randell, B., and Russell, L. J., 1964, *Algol 60 Implementation,* Academic Press, New York.

Exercises

1. Evaluate the following arithmetic expressions directly from the various forms of Polish notation:
   (a) $+ x + x + 1\ 2\ 3\ 4\ 5\ 6$
   (b) $x + x\ 2 + 3\ 4\ 5\ 6$
   (c) $-6 - 5 - 4 - 3 - 2\ 1$
   (d) $6\ 5\ 4\ 3\ 2\ 1 - - - - -$
   (e) $- - - - - 6\ 5\ 4\ 3\ 2\ 1$
   (f) $1\ 2\ 3 + 4\ 5 + 6 \times \times +$

2. Translate each of the examples in Question 1 into conventional notation and re-evaluate.

3. Draw the tree structure for each of the following expressions:
   (a) $w \& (x \& (y \& z))$
   (b) $(w \& (x \& y)) \& z$
   (c) $(x \& y \vee \bar{y} \& \bar{z}) \not\equiv (x \rightarrow (\bar{y} \rightarrow \bar{z}))$
   (d) $\overline{x \vee y \vee z} \& \overline{\bar{x} \vee \bar{y} \vee \bar{z}}$

4. Draw the tree structure for each of the following Polish strings:
   (a) $w\,x \rightarrow y \vee z \,\&$
   (b) $\& \,\& \,\& \,s\,t \,\& \,u\,v \,\& \,\& \,w\,x \,\& \,y\,z$
   (c) $N \& N \vee w\,x\,N \vee y\,z$
   (d) $w\,N\,x\,N\,y\,z \vee N \& N \& N$

5. Write out the four standard forms of Polish notation for each of the expressions:
   (a) $x \vee y \vee z \& \bar{x} \vee \bar{y} \vee \bar{z}$
   (b) $(x \vee y \vee z) \& (\bar{x} \vee \bar{y} \vee \bar{z})$
   (c) $x \& y \vee y \& z$
   (d) $\bar{w} \& \bar{x} \& \overline{\bar{y} \& \bar{z}} \& x \& y \& z$

6. Convert these forward Polish expressions into reverse Polish, keeping the same order of variables:
   (a) $\& \,N\,w \,\& \,N\,x \,\& \,N\,y\,N\,z$
   (b) $\& \,\& \,\& \,N\,w\,N\,x\,N\,y\,N\,z$
   (c) $\& \vee \vee \vee w\,x\,y\,z \,\& \,z \,\& \,y \,\& \,x\,w$
   (d) $\rightarrow \rightarrow w\,x \rightarrow y\,z$

7. Rewrite the Polish expressions of Question 6 as conventional Boolean expressions.

8. Simplify the following expressions, giving the answer in early operator reverse Polish notation:
   (a) $\vee \vee \& x\,N\,y \& N\,x\,N\,y \& N\,x\,y$
   (b) $\vee \not\equiv x\,y \equiv y\,z$
   (c) $K\,K\,H\,x\,y\,C\,y\,z\,C\,y\,N\,z$

Chapter 8
Designation Numbers

Introduction

The use of binary numbers to represent propositional functions in a form known as designation numbers provides a convenient mathematical technique to manipulate and simplify Boolean expressions. They were devised by R. S. Ledley and a full treatment may be found in Ledley (1960).

A unique designation number is associated with every propositional variable or expression. Any manipulations or simplifications required are carried out on the designation numbers and this is then transformed back into a Boolean expression to give the required solution.

The Basis

Designation numbers are first assigned to each of the elementary variables x, y, z, etc., that occur in the Boolean expressions under consideration. They are based upon the rows that appear in the truth table. The truth table is rotated through

90° so that all the alternative values of the elementary elements appear as columns instead of rows. Thus the truth table shown below

Row No.	x	y	z
0	0	0	0
1	0	0	1
2	0	1	0
3	0	1	1
4	1	0	0
5	1	0	1
6	1	1	0
7	1	1	1

is first rotated to give

$$z \quad 0\,1\,0\,1 \quad 0\,1\,0\,1$$
$$y \quad 0\,0\,1\,1 \quad 0\,0\,1\,1$$
$$x \quad 0\,0\,0\,0 \quad 1\,1\,1\,1$$

column number $\qquad 0\,1\,2\,3 \quad 4\,5\,6\,7$

The numbers opposite each of x, y, z are the designation numbers, denoted by the symbol #, of x, y, z.

$$\#z = 0\,1\,0\,1 \quad 0\,1\,0\,1$$
$$\#y = 0\,0\,1\,1 \quad 0\,0\,1\,1$$
$$\#x = 0\,0\,0\,0 \quad 1\,1\,1\,1$$

column $\quad 0\,1\,2\,3 \quad 4\,5\,6\,7$

The assignment of designation numbers to the elementary variables is known as the *basis*. It should be noted that, as chosen above, the binary numbers formed by the columns, read from bottom to top, give the column number. This particular assignment will be known as the *standard* basis and be denoted by $b[z, y, x]$.

Any other permutation of the columns could equally well be chosen as a basis, giving in this instance $8.7.6.5.4.3.2.1$ or $8!$ different choices. It is generally convenient to use the standard basis to save confusion and this convention will be adhered to unless otherwise stated.

The number of columns in the standard basis depends on the number of variables under consideration. For n variables, 2^n columns are required for a complete basis and would be numbered from 0 to $2^n - 1$. Thus, with 4 variables, there are 16 columns and each designation number consists of

16 binary digits. The sets of standard bases for 2, 3, and 4 variables are as follows:

$\#z = 0\,1\,0\,1$	$0\,1\,0\,1$	$0\,1\,0\,1$	$0\,1\,0\,1$
$\#y = 0\,0\,1\,1$	$0\,0\,1\,1$	$0\,0\,1\,1$	$0\,0\,1\,1$
$\#x = 0\,0\,0\,0$	$1\,1\,1\,1$	$0\,0\,0\,0$	$1\,1\,1\,1$
$\#w = 0\,0\,0\,0$	$0\,0\,0\,0$	$1\,1\,1\,1$	$1\,1\,1\,1$

The standard basis is very simple to write out. The first row consists of the numbers 0, 1 alternately. The second has two 0s followed by two 1s, the third four 0s followed by four 1s, then sets of eight, sixteen, etc. In each case the pattern is continued to give the full 2^n columns.

The order of the rows in the standard basis b$[z, y, x, w]$ with z at the top rather than b$[w, x, y, z]$ with w at the top has been chosen since it is then very easy to extend the basis downwards when more variables are required.

Designation Numbers of Expressions

The designation number of any Boolean function is obtained by first assigning the standard basis to the elementary variables and then performing the indicated logical operations on the designation numbers, using the normal laws of Boolean algebra applied to each column in turn.

Thus to find the designation number of $y \cdot z$ set up the standard basis b$[z, y]$

$$\#z = 0101$$
$$\#y = 0011$$
$$\overline{\#y \cdot z = 0001}$$

The result is obtained by taking the logical product of the pairs of digits in each column of the two designation numbers. The method can also be applied for *any* of the propositional connectives, using the definitions as provided by the truth tables. Thus for $f = (\overline{x} \to y) \,\&\, (y \not\equiv z)$

$$\#z = 0101 \quad 0101$$
$$\#y = 0011 \quad 0011$$
$$\#x = 0000 \quad 1111$$
$$\overline{\#\overline{x} = 1111 \quad 0000}$$
$$\#(\overline{x} \to y) = 0011 \quad 1111$$
$$\#(y \not\equiv z) = 0110 \quad 0110$$
$$\overline{\#f = 0010 \quad 0110}$$

Properties of Designation Numbers

The most important property is that

$$\# f = \# g$$

is and only is

$$f = g$$

Thus, to show that two Boolean functions f and g are equiveridic, it is only necessary to calculate the designation numbers of f and g. If they are equal, the functions are equiveridic.

The universally false (O) and universally true (I) functions have the designation numbers

$$\# O = 0000 \quad 0000 \quad ----, \text{ i.e. all zeros}$$
$$\# I = 1111 \quad 1111 \quad ----, \text{ i.e. all ones}$$

Thus the designation number of any tautology consists entirely of 1s.

The truth or falsity of a logical implication may be determined immediately by inspection of the designation numbers. Recall that f logically implies g ($f \Rightarrow g$) if and only if ($f \rightarrow g$) is a tautology.

To show that $y \cdot z \Rightarrow y + z$, use the standard basis $b[z, y]$. Then

$$\# (y \cdot z) = 0001$$
$$\# (y + z) = 0111$$
$$\overline{\# [(y \cdot z) \rightarrow (y + z)] = 1111}$$

i.e. a tautology. Hence $y \cdot z \Rightarrow y + z$. However, this can be seen immediately from the designation numbers of the two expressions: $f \Rightarrow g$ if and only if $\# g$ has units in at least those positions which correspond to the units of $\# f$. $\# (y \cdot z)$ has units only in column 3. $\# (y + z)$ also has a unit in this column, hence

$$y \cdot z \Rightarrow y + z$$

Special Disjunctive Normal Form

The special disjunctive normal form consists of a sum of products, each product containing *all* the variables either naturally or negated. Thus for three variables x, y, z, each term is of the form $\bar{x} \cdot \bar{y} \cdot \bar{z}$, $\bar{x} \cdot \bar{y} \cdot z$, etc. Since the main interest is now on the overall expression rather than the emphasis of a particular operation, it will be convenient to drop the Boolean product sign (.) as in ordinary multiplication and to write these terms as $\bar{x} \bar{y} \bar{z}$, $\bar{x} \bar{y} z$, etc. Thus using the standard basis $b[z, y, x]$, consider the designation numbers of each of the 2^3 *elementary product* terms in all three variables,

$$\#(\bar{x}\,\bar{y}\,\bar{z}) = 1000 \quad 0000$$
$$\#(\bar{x}\,\bar{y}\,z) = 0100 \quad 0000$$
$$\#(\bar{x}\,y\,\bar{z}) = 0010 \quad 0000$$
$$\#(\bar{x}\,y\,z) = 0001 \quad 0000$$
$$\#(x\,\bar{y}\,\bar{z}) = 0000 \quad 1000$$
$$\#(x\,\bar{y}\,z) = 0000 \quad 0100$$
$$\#(x\,y\,\bar{z}) = 0000 \quad 0010$$
$$\#(x\,y\,z) = 0000 \quad 0001$$

If in each of the terms $\bar{x}\,\bar{y}\,\bar{z}$, etc., the value 0 is substituted for a negated variable and 1 for a natural variable, three-digit binary numbers 000, 001, 010, etc., are obtained for each row of the above array. Converted to decimal notation, this gives the row numbers 0, 1, 2, . . ., 7 and it is a convenient shorthand to use these decimal numbers to refer to each row. Thus, for instance, the product term $x\,\bar{y}\,z$ with designation number 0000   0100 occurs in row 5 and will be denoted by P_5.

Each designation number contains exactly one unit term and this occurs in column 0 for the term with decimal number 0, column 1 for the term with decimal number 1, etc. Thus it is possible to say immediately that $\#x\,\bar{y}\,z$ has a 1 in column 101 = 5 and zeros elsewhere. This will only hold true if the literals in each term are ordered in the same fashion as the rows of the standard basis (from bottom to top). If the variables in a Boolean expression appear in the order a, c, b, and d, say, the basis $b[d, b, c, a]$ must be used.

The special disjunctive normal form consists of the logical sum of product terms of this kind, each term contributing therefore a single 1 to the designation number. For example

$$0010 \quad 0110$$

is the result of the logical sum of the designation numbers with units in columns 2, 5, and 6,

$$\#(P_2 = P_{010} = \bar{x}\,y\,\bar{z}) = 0010 \quad 0000$$
$$\#(P_5 = P_{101} = x\,\bar{y}\,z) = 0000 \quad 0100$$
$$\#(P_6 = P_{110} = x\,y\,\bar{z}) = 0000 \quad 0010$$

Hence $\#(\bar{x}\,y\,\bar{z} + x\,\bar{y}\,z + x\,y\,\bar{z}) = 0010 \quad 0110$

Provided that the standard basis has been used as in the above examples, the corresponding terms for each unit can be written down immediately by converting back from decimal to binary and hence to the term itself. However, if a different basis has been used, the terms must be determined from an inspection of the actual basis.

As a further example, suppose a basis $b[z, y, b, a]$ has been chosen and an expression yields the designation number

$$1000 \quad 0001 \quad 1000 \quad 0001$$

then columns 0, 7, 8, 15 contain units. In decimal suffix notation, the special disjunctive normal form is

$$P_0 + P_7 + P_8 + P_{15}$$

and in binary suffix form this is

$$P_{0000} + P_{0111} + P_{1000} + P_{1111}$$

giving

$$\bar{a}\bar{b}\bar{y}\bar{z} + \bar{a}byz + a\bar{b}\bar{y}\bar{z} + abyz$$

Special Conjunctive Normal Form

The procedure is the exact dual of the special disjunctive normal form. A consideration of each of the factors or elementary sums $\bar{x} + \bar{y} + \bar{z}$, $\bar{x} + \bar{y} + z$, etc., of a special conjunctive normal form shows that each designation number contains exactly one zero:

$$
\begin{aligned}
\#(x + y + z) &= 0111 \quad 1111 \\
\#(x + y + \bar{z}) &= 1011 \quad 1111 \\
\#(x + \bar{y} + z) &= 1101 \quad 1111 \\
\#(x + \bar{y} + \bar{z}) &= 1110 \quad 1111 \\
\#(\bar{x} + y + z) &= 1111 \quad 0111 \\
\#(\bar{x} + y + \bar{z}) &= 1111 \quad 1011 \\
\#(\bar{x} + \bar{y} + z) &= 1111 \quad 1101 \\
\#(\bar{x} + \bar{y} + \bar{z}) &= 1111 \quad 1110
\end{aligned}
$$

The product of a number of elementary sums will give a zero in the result corresponding to the positions of zeros in the elementary sums. However, a zero in column 1, say, obtained from row 1, which will be denoted by S_1, gives the sum $x + y + \bar{z}$ (i.e. 110). This is the negation of binary 001 corresponding to decimal 1. Thus the designation number

$$0010 \quad 0110$$

requires zeros in columns 0, 1, 3, 4, 7 giving a special conjunctive normal form

$$(\text{row } 0) \cdot (\text{row } 1) \cdot (\text{row } 3) \cdot (\text{row } 4) \cdot (\text{row } 7)$$

or

$$S_0 \cdot S_1 \cdot S_3 \cdot S_4 \cdot S_7$$

that is

$$(x + y + z)(x + y + \bar{z})(x + \bar{y} + \bar{z})(\bar{x} + y + z)(\bar{x} + \bar{y} + \bar{z})$$

Decimal Notation

This decimal notation introduced can be used as a useful shorthand method of expressing normal forms and in some instances the calculations may be performe on the decimal numbers themselves.

Any elementary sum factor S_i may be converted to an elementary product term P_i and vice versa by the relations

$$P_i = \overline{S_i} \qquad \text{and} \qquad S_i = \overline{P_i}$$

As a simple example, consider the conversion of the expression in conjunctive normal form in four variables

$$f = S_1 \cdot S_5 \cdot S_9 \cdot S_{13}$$

into disjunctive normal form. Take the negation of f, giving

$$\bar{f} = \overline{S_1 \cdot S_5 \cdot S_9 \cdot S_{13}}$$
$$= \overline{S_1} + \overline{S_5} + \overline{S_9} + \overline{S_{13}}$$
$$= P_1 + P_5 + P_9 + P_{13}$$

The expression for f, in special disjunctive normal form, is now given by all those terms of the complete disjunctive normal form which are *absent* from the expression for $\bar{f}$ in special disjunctive normal form, namely

$$f = P_0 + P_2 + P_3 + P_4 + P_6 + P_7 + P_8 + P_{10} + P_{11} + P_{12} + P_{14} + P_{15}$$

Designation Numbers of Other Normal Forms

The special normal forms include *all* the n literals in each product term or sum factor. In a general disjunctive normal form, some of the terms will contain fewer than n products, e.g. in the three-variable expression $\bar{x}y\bar{z} + xz$, xz is a two-term product. The designation number of such a product can be found by taking the logical product of the designation numbers of x and z. With the standard three-variable basis $b[z, y, x]$ this is

$$
\begin{array}{rll}
\#z = & 0101 & 0101 \\
\#x = & 0000 & 1111 \\
\hline
\#xz = & 0000 & 0101 \\
\end{array}
$$

Thus the designation number contains two 1s.

In general the designation number of a p-term product with an n-variable basis will contain 2^{n-p} units. The positions of these units can also be obtained by Boolean algebra since

$$x\,z = x(\bar{y} + y)\,z$$
$$= x\,\bar{y}\,z + x\,y\,z$$
$$= P_5 + P_7$$

so that there are 1s in columns 5 and 7. This indicates a method of writing out the designation number of such product terms by inspection. Any variable that does not occur in the product term must occur in both natural and negated forms in the expansion into special normal form. Thus both possible values of that binary position should be included. For example with a four-variable basis $b[z, y, x, w]$, the two-variable term $\bar{x}y$ must take both values of the 2^0 and 2^3 power positions with the binary numbers

$$0\;01\;0; \quad 0\;01\;1; \quad 1\;01\;0; \quad 1\;01\;1$$

so that the product may be written in special disjunctive normal form as

$$\bar{x}\,y = P_2 + P_3 + P_{10} + P_{11}$$

with a designation number 0011 0000 0011 0000 containing units in columns 2, 3, 10, and 11.

The designation number of conjunctive normal forms may be found by a similar procedure. With the same basis, $w + \bar{y} + \bar{z}$ is obtained as follows. Only the variable x is absent with power position 2^2 giving the two sum terms represented by

$$S_{0011} \cdot S_{0111}$$
$$= S_3 \cdot S_7$$

The designation number therefore has zeros in columns 3 and column 7, i.e.

$$\# w + \bar{y} + \bar{z} = 1110 \quad 1110 \quad 1111 \quad 1111$$

References and Further Reading

Chu, Y., 1962, *Digital Computer Design Fundamentals*, McGraw-Hill, New York.

Ledley, R. S., 1960, *Digital Computer and Control Engineering*, McGraw-Hill, New York.

Exercises

1. Write out the following designation numbers:
   (a) $\# v$ for the six-variable standard basis $b[z, y, x, w, v, u]$
   (b) $\# a$ for the basis $b[a, b, c]$
   (c) $\# x$ for the basis $b[z, x, w, y]$

2. By determining the designation numbers of both sides of the equations, justify the following theorems of Boolean algebra:
   (a) $y + y \cdot z = y$
   (b) $x \cdot (y + z) = x \cdot y + x \cdot z$
   (c) $\overline{y \cdot z} = \bar{y} + \bar{z}$
   (d) $y + \bar{y} \cdot z = y + z$

3. From the designation numbers obtained in Question 2, write out the two special normal forms for each of the expressions.

4. Determine, by using designation numbers, which of the following compound propositions are tautologies, contradictions, or neither:
   (a) $x \,\&\, \bar{x} \lor y \,\&\, \bar{y} \lor z \,\&\, \bar{z}$
   (b) $[(x \to y) \,\&\, x] \to y$
   (c) $(\bar{x} \lor y) \,\&\, (\bar{z} \lor y) \to (\bar{x} \lor z)$
   (d) $[(\bar{x} \lor y) \,\&\, (\bar{y} \lor z) \,\&\, x] \to z$
   (e) $(x \to y) \to (x \to y)$
   (f) $((x \to y) \to x) \to y$

5. Demonstrate the validity or otherwise of the following logical implications:
   (a) $(a \to b) \Rightarrow (a \to b)$
   (b) $(p \to q) \,\&\, (q \to r) \Rightarrow (p \to r)$
   (c) $(x \to y) \,\&\, (z \to y) \Rightarrow (x \to z)$
   (d) $(x \equiv y) \Rightarrow (x \to y)$
   (e) $p \,\&\, \bar{p} \Rightarrow z$

6. Find the special disjunctive normal form for each of the expressions:
   (a) $(x \to y) \cdot (y \to z)$
   (b) $[(x \equiv y) \cdot (z \to \bar{y})] \downarrow [(\bar{x} \not\equiv z) | (\bar{x} \leftarrow y)]$
   (c) $(p \downarrow q \downarrow r) | (\bar{p} \downarrow \bar{q} \downarrow \bar{r}) | (p \downarrow \bar{q} \downarrow r)$
   (d) $(a \to b) \equiv (\bar{b} \to \bar{a})$

7. Use the basis $b[x, y, z]$ to find designation numbers with respect to this basis of the Boolean expressions:
   (a) $x \downarrow y \downarrow z$
   (b) $(x \downarrow y) \downarrow z$
   (c) $x \downarrow (y \downarrow z)$

Hence write out the special disjunctive normal forms.

8. Write out the Boolean expression for each of the following logical sums or products which are expressed with respect to the given bases.
   (a) $\quad P_1 + P_3,\ b[z, y]$
   (b) $\quad P_{63},\ b[z, y, x, w, v, u]$
   (c) $P_0 + P_1 + P_{14},\ b[a, b, x, y, z]$
   (d) $\quad S_2 . S_0,\ b[z, y]$
   (e) $\quad S_{63},\ b[z, y, x, w, v, u]$
   (f) $S_{31} . S_{30} . S_{17},\ b[a, b, x, y, z]$

9. Convert each of the following expressions into the opposite type of special normal form by using decimal notation, i.e. convert special disjunctive normal form to special conjunctive normal form and vice versa.
   (a) $(y + z)(\bar{y} + z)$
   (b) $y z + \bar{y} z$
   (c) $P_0 + P_1 + P_{14} + P_{15}$
   (d) $S_1 . S_2 . S_{12} . S_{13}$

10. Simplify the expression in four variables

$$P_2 . S_2 + (P_5 + S_5) . (P_3 + \bar{S}_3)$$

Chapter 9
Karnaugh Maps

Simplification

Using truth tables or designation numbers the special disjunctive or conjunctive normal forms can be obtained for any Boolean or propositional function. Logic or switching circuits can easily be obtained from these results but they will generally use many more logical elements than are strictly necessary. It is therefore desirable to simplify the functions before transferring to the logic circuit. The techniques of Boolean algebra give one possible method of simplifying the expressions but are not particularly easy to apply, nor is it possible to recognize the simplest form of the expression when it is reached. This is particularly true when functions of several variables are involved.

Systematic methods are therefore required to find the simplest expression for any given Boolean function. But first it is necessary to decide what is meant by 'simplest'. Many criteria could be used, examples being as follows.

(i) The number of occurrences of literals in the expression, e.g. there are four in the expression $(\bar{x} + y) \cdot (x + \bar{y})$.

(ii) The number of operations that must be performed. There are five in the above example if negation is counted as an operation or three otherwise.

There are many other factors to be considered when the problem is to obtain a physical circuit and not just a simplified Boolean expression. Some of these are listed below.

(i) Total number of transistors used.

(ii)   The number of logical levels.

(iii) Maximum fan-in, i.e. restrictions on the number of inputs to a gate.

(iv) Maximum fan-out, i.e. restrictions on the number of outputs of a gate.

(v)   Comparative cost of different types of gates, AND, OR, NOR, etc.

(vi) Power dissipation.

(vii) Reliability.

(viii) Whether the circuit is fail-safe.

If relay switching circuits are to be used, then some of the considerations are the following.

(i)   Number of relay contacts.

(ii) Number of springs.

(iii) Types of relays available; single or double throw, number of poles, etc.

The number of switching functions of n variables is 2^{2^n}. This increases very rapidly as a function of n and is shown in Table 9.1. These numbers are far too

Table 9.1. Number of switching functions of n variables

No. of variables	No. of functions
n	2^{2^n}
1	4
2	16
3	256
4	65 536
5	4 294 967 296

large to investigate all possible circuits in order to minimize a Boolean function taking all the above factors into account. However, techniques are well-developed for determining the simplest normal form in the sense of the minimum number of occurrences of variables and it is these methods that will mainly be considered now and also in Chapter 10.

Any further reduction of the circuit must then depend on the skill and knowledge of the designer, taking into account the physical characteristics of the circuit elements to be used. This will be particularly true of relay switching networks, as the methods of Boolean algebra lead to series-parallel circuits obtained directly from a sum of products or product of sums. The class of circuits known as bridge circuits as illustrated in Figure 9.1 cannot easily be

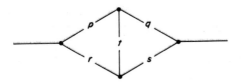

Figure 9.1

synthesized although they can be readily analysed by Boolean algebra. There is no general method of determining whether a simpler bridge network exists that is equivalent to any given series-parallel circuit but some useful techniques for obtaining bridge circuits will be given in Chapter 11.

The Karnaugh Map for Two Variables

The Karnaugh map is a stylized version of a Venn diagram and is probably the easiest method of simplifying Boolean expressions in normal form for up to four or five variables (Karnaugh, 1953). It can be extended to six or more variables but much of the simplicity of the method is then lost.

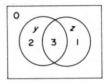

Figure 9.2

Consider, first of all, functions of two variables y, z illustrated in the Venn diagram of Figure 9.2. The four minimal regions are represented by each of the possible product terms

$$\text{area } 0 = P_0 = P_{00} = \bar{y}\,\bar{z}$$
$$\text{area } 1 = P_1 = P_{01} = \bar{y}\,z$$
$$\text{area } 2 = P_2 = P_{10} = y\,\bar{z}$$
$$\text{area } 3 = P_3 = P_{11} = y\,z$$

The other expressions in two variables that occur are

$$y = y(z + \bar{z}) = y\,z + y\,\bar{z} = P_3 + P_2$$
$$z = (y + \bar{y})z + y\,z + \bar{y}\,z = P_3 + P_1$$

and

$$y = \overline{P_3 + P_2} = P_0 + P_1$$
$$\bar{z} = \overline{P_3 + P_1} = P_0 + P_2$$

Each of these functions represents the sum or union of two of the minimal regions. Finally, the always true function I is the union of all four regions $0 + 1 + 2 + 3$.

A disjunctive normal form in two variables consists of the logical sum of selections of these nine possibilities

$$\bar{y}\,\bar{z},\ \ \bar{y}\,z,\ \ y\,\bar{z},\ \ y\,z,\ \ y,\ \ z,\ \ \bar{y},\ \ \bar{z},\ \ I$$

If the Venn diagram is redrawn in the form of Figure 9.3 it is known as a Karnaugh map. Each individual square corresponds to one of the original minimal regions and represents

$$\bar{y}\,\bar{z}\,(0), \quad \bar{y}\,z\,(1), \quad y\,\bar{z}\,(2), \quad \bar{y}\,\bar{z}\,(3)$$

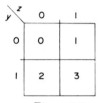

Figure 9.3

The individual truth values of y and z are indicated along the edges of the map. The other possibilities

$y = 2 + 3$	corresponds to the whole row $y = 1$
$z = 1 + 3$	corresponds to the whole column $z = 1$
$\bar{y} = 0 + 1$	corresponds to the whole row $y = 0$
$\bar{z} = 0 + 2$	corresponds to the whole column $z = 0$
$I = 0 + 1 + 2 + 3$	corresponds to the whole diagram

To simplify a Boolean expression given in special disjunctive normal form, place a 1 in each square that appears in the expression. The Karnaugh map for the function

$$f = y\,\bar{z} + \bar{y}\,z + y\,z$$

is shown in Figure 9.4.

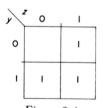

Figure 9.4

Since any two 1s in the same row or column can be described by one variable for both values of the other variable, the combination of two such 1s requires only a single variable. Thus to obtain the simplified form an attempt must be made to combine the 1s that appear in the map in as large a section as possible. If all four squares contain a 1, the function is universally true, I. Otherwise, *draw loops around all possible horizontal or vertical pairs attempting to enclose all the 1s in loops. Overlapping is permitted.*

Drawing the loops in Figure 9.4 leads to Figure 9.5. Two loops are required, one in the column $z = 1$, the other in the row $y = 1$. Hence the simplified expression is

$$f = y + z$$

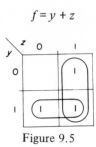

Figure 9.5

When drawing the map for a disjunctive normal form rather than a special disjunctive normal form, if a term with only one literal such as y or $\bar{z}$ occurs, then 1s must be placed in both the appropriate squares. If a square already contains a 1, then no additional mark need be made if it occurs again.

The Karnaugh Map for Three or Four Variables

The Venn diagram for four variables may be drawn as in Figure 9.6. There are now 16 minimal regions. It can be seen that it is quite a tricky task to draw it

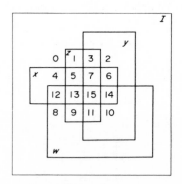

Figure 9.6

correctly and obtain the right number of these minimal areas. The four-variable Venn diagram can be rearranged to give the Karnaugh map of Figure 9.7. With this number of variables in particular, the advantages of the Karnaugh map representation became very apparent. The decimal numbers have been included to make reference easier. They are not used in actual minimizations. Note particularly the ordering of the columns and rows. The order is 00, 01, 11, 10 *with only one of the two digits changing each time. It is not in binary order.*

Provided the sequence of numbering is such that one digit is altered between adjacent rows or columns, any other order of variables is permitted but the sequence given above is almost universally adopted.

wx \ yz	00	01	11	10
00	0	1	3	2
01	4·	5	7	6
11	12	13	15	14
10	8	9	11	10

Figure 9.7

Not only is the diagram very much easier to draw and understand, but the particular method of ordering the rows and columns permits immediate recognition of terms that can be simplified. It relies on the Boolean equation

$$\bar{y}z + yz = (\bar{y} + y).z = z$$

The two product terms $\bar{y}z$, yz would be represented by columns 01, 11 respectively in the Karnaugh map which are necessarily adjacent in the map as they differ in one digit only. Indeed, as will be seen below, any terms that are represented by adjacent squares (horizontally or vertically, but not diagonally) can always be simplified.

A single square represents the occurrence of a term containing all four literals, e.g. $wxyz$ (= 1111 = 15). A pair of horizontal or vertical squares always represents a term with only three literals. Owing to the ordering of the columns and rows, two adjacent squares only differ in *one* of the four digits. Thus the pair of squares 0101 (5) and 1101 (13) differ in the digit that represents w. Together they represent both possible values of w, i.e.

$$\bar{w}x\bar{y}z + wx\bar{y}z$$
$$= (\bar{w} + w).x\bar{y}z$$
$$= x\bar{y}z$$

Similarly the terms $wx\bar{z}$ is represented by the squares 1100 (12) and 1110 (14) for both values of the y variable.

Groups of four squares together have *two* digits that take both 0 and 1 as values and therefore represent the product of two variables. For example, xz consists of the squares 0101, 0111, 1101, 1111 (Figure 9.8). Similarly eight squares represent a single variable, and the whole sixteen squares represent I.

To carry out a simplification the appropriate squares are marked on the Karnaugh map. An attempt is then made to group the marked squares into larger squares, rectangles, rows, columns, or adjacent horizontal or vertical pairs, the number of squares in any group being a power of 2. Overlapping is, of course, still permitted. For the simplest disjunctive normal form, the 1s in the map should be grouped together to form the fewest possible loops, each loop enclosing the largest number of permitted squares. Diagonal grouping is

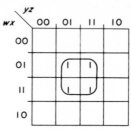

Figure 9.8

not permitted. However, owing to the ordering of columns and rows, the first square in any given row (e.g. 0100) and the last square in that row (0110) differ in only one digit and can be considered to be adjacent. The same reasoning applies to columns. Thus end-around looping is permitted. In particular, the four corner squares 0000, 1000, 1010, 0010 may be looped as in Figure 9.9 to give $\bar{x}\,\bar{z}$.

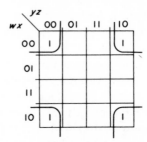

Figure 9.9

As an example, the function

$$f = \overline{w}\,\overline{x}\,\overline{y} + \overline{w}\,\overline{x}\,z + \overline{w}\,\overline{x}\,y + \overline{w}\,\overline{x}\,y\,z + w\,\overline{x}\,y\,z + w\,\overline{x}\,y\,\overline{z}$$

will be simplified. The Karnaugh map for this function is shown in Figure 9.10.

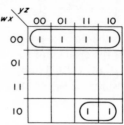

Figure 9.10

If the 1s are looped as in Figure 9.10, the simplified expression is

$$f = \overline{w}\,\overline{x} + w\,\overline{x}\,y$$

This, however, is not the minimum expression as the loop with two squares in the bottom right-hand corner can be overlapped with two in the top right-hand

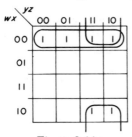

Figure 9.11

corner to give a loop of 4 as in Figure 9.11. This gives the correct minimum disjunctive normal form

$$f = \overline{w}\,\overline{x} + \overline{x}\,y$$

As can easily be seen in this instance, this is not an *absolute* minimum form, since it can be factorized to yield

$$f = \overline{x}\,.\,(\overline{w} + y)$$

with only three occurrences of variables. The Karnaugh map method gives the *appropriate normal* form with the fewest occurrences of literals.

A warning should be given concerning the best arrangement of looping. Do not rush in and immediately loop together the 1s that form the largest loop that can be obtained. Consider the Karnaugh map in Figure 9.12, in which the

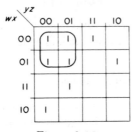

Figure 9.12

group of four in the top left-hand corner has immediately been looped. Each of the remaining four squares can be grouped in loops of two as in Figure 9.13, giving the function

$$\overline{w}\,y + \overline{w}\,\overline{x}\,z + \overline{w}\,x\,\overline{y} + x\,\overline{y}\,z + \overline{x}\,\overline{y}\,\overline{z}$$

But the four two-square loops in fact cover all the marked squares giving the simpler expression

$$\overline{w}\,\overline{x}\,z + \overline{w}\,x\,\overline{y} + x\,\overline{y}\,z + \overline{x}\,\overline{y}\,\overline{z}$$

so that the term $\overline{w}\,y$ (the four-square loop) is not required at all.

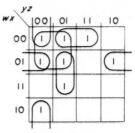

Figure 9.13

Therefore, although the largest possible loops are required, the procedure should be carried out in the following sequence, stopping when *all* the marked squares are looped.

(i)   Loop all 1s that cannot combine with any other 1.

(ii) Loop all 1s that will combine in a loop of 2 but not of 4.

(iii) Loop all 1s that will combine in a loop of 4 but not of 8, etc.

For a Karnaugh map with three variables, the bottom two rows may be removed as in Figure 9.14. The procedure is the same as the four-variable case.

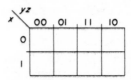

Figure 9.14

The special disjunctive normal form may also be obtained from a Karnaugh map by reading off each marked square individually, with no looping.

Karnaugh Map from Designation Number

When the Boolean function to be minimized is given in the form of a designation number, it is not necessary to go through the intermediate stage of writing out the special disjunctive normal form. It is quite easy to proceed direct from the designation number to the Karnaugh map.

In Chapter 8, the designation number for the function

$$f = (\overline{x} \rightarrow y)\ \&\ (y \not\equiv z)$$

was obtained as $\#f = 0010\ \ 0110$. To simplify this, draw a three-variable

Karnaugh map, Figure 9.15. The designation number has 1s in columns 2, 5, 6 or in binary 010, 101, 110. Thus a 1 should be placed in these squares on the Karnaugh map. Reading off the answer from the map gives

$$f = y\bar{z} + x\bar{y}z$$

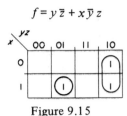

Figure 9.15

Simplification of Conjunctive Normal Form

This is done in a similar manner to the disjunctive normal form by using the principle of duality. Firstly, any square in the Karnaugh map that is not a 1 may be marked as a zero. Thus to obtain a conjunctive normal form from an expression given as a disjunctive form, simply consider all the squares without 1s. Thus for the above example of $f = (\bar{x} \rightarrow y)$ & $(y \not\equiv z)$, the Karnaugh map for the conjunctive normal form is as in Figure 9.16. This time the 0s are looped. To read off the answer in conjunctive form, the dual of each looped group must be taken, so that if the x value is 0 take x, and if it is 1 take $\bar{x}$. Thus the loop of

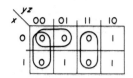

Figure 9.16

two 0s in the left-hand column of Figure 9.16 represents $y + z$ and the complete expression simplifies to the conjunctive normal form

$$f = (y + z)(x + y)(\bar{y} + \bar{z})$$

When proceeding from a conjunctive form, a zero should be placed on the map in each square corresponding to the dual of the sum term, e.g. for the factor $\bar{w} + x + \bar{y} + z$ a 0 is placed in the square 1010. Commencing with a designation number, either the 1s can be inserted from the units of the number or 0s from the zeros. The quickest procedure is to use whichever has the fewer appearances.

Use of Don't Care Conditions

It frequently happens that the designer of a piece of equipment does not mind what result a circuit gives for some particular combination of values of the variables. This may be because he knows that this input condition cannot occur in practice. These combinations are known as '*don't care*' conditions. The value of 0 or 1 can be used and the value is assumed that leads to the greatest simplification. The 'don't care' condition can be indicated in a designation number by the symbol X. For the two variable designation number 100X, this indicates a function $\overline{y}\,\overline{z}$ with a don't care condition $y\,z$. When the Karnaugh map is drawn, the square (or squares) corresponding to the don't care conditions will also be marked with X. This can then be used as either 0 or 1 when the loops are drawn.

The expression

$$f = \overline{w}\,x + x\,\overline{y} + x\,y\,z$$

will be simplified using the don't care condition $w\,y\,\overline{z}$. The Karnaugh map is drawn in Figure 9.17. The don't cares give two squares which can be chosen at

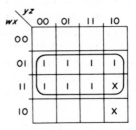

Figure 9.17

will. If $w\,x\,y\,\overline{z}$ is taken as 1, this gives a group of 8 that can be looped. Hence

$$f = x$$

will produce the desired function.

Karnaugh Maps for More Than Four Variables

The concept of simplification by Karnaugh maps may be extended to five, six, or even more variables. But the simplicity of the method is no longer present, as it is no longer possible to see adjacent squares immediately. Five variables can be accommodated by placing two four-variable maps side by side. They should of course be above one another, so that looping can also be made vertically. This could be done with a sheet of transparent plastic.

For an example of the use of a five-variable map, consider the simplification of the expression

$$f = \overline{\overline{y} \to v\,w\,\overline{y}} + [(w\,\overline{x}) \equiv (x + y)] + v\,w\,x\,y\,z$$

This is an expression in five variables and the standard basis $b[z, y, x, w, v]$ will be used to obtain its designation number.

$$
\begin{array}{llll}
\#z & = 01010101 & 01010101 & 01010101 & 01010101 \\
\#y & = 00110011 & 00110011 & 00110011 & 00110011 \\
\#x & = 00001111 & 00001111 & 00001111 & 00001111 \\
\#w & = 00000000 & 11111111 & 00000000 & 11111111 \\
\#v & = 00000000 & 00000000 & 11111111 & 11111111 \\
\end{array}
$$

$$
\begin{array}{llll}
\#\bar{y} & = 11001100 & 11001100 & 11001100 & 11001100 \\
\#vw\bar{y} & = 00000000 & 00000000 & 00000000 & 11001100 \\
\#\bar{y} \to vw\bar{y} & = 00110011 & 00110011 & 00110011 & 11111111 \\
\#\overline{\bar{y} \to vw\bar{y}} & = 11001100 & 11001100 & 11001100 & 00000000 \\
\#w\bar{x} & = 00000000 & 11110000 & 00000000 & 11110000 \\
\#[(w\bar{x}) \equiv (x + y)] & = 11000000 & 00110000 & 11000000 & 00110000 \\
\#vwxyz & = 00000000 & 00000000 & 00000000 & 00000001 \\
\#f & = 11001100 & 11111100 & 11001100 & 00110001 \\
\end{array}
$$

The Karnaugh map for this function is shown in Figure 9.18. Loops that include both values of the variable v have been shown in solid lines and other

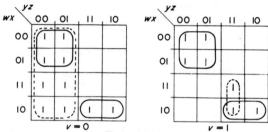

Figure 9.18

loops in dashed lines. From the map, the simplest disjunctive normal form is

$$f = \bar{v}\bar{y} + \bar{w}\bar{y} + w\bar{x}z + vwyz$$

An alternative method for five variables is to divide each square into two halves as in Figure 9.19. The groupings cannot normally be made as loops but must be indicated by some kind of coded mark.

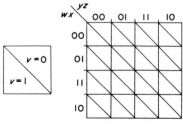

Figure 9.19

For six variables, four of the basic four-variable maps may be laid out in a square or two diagrams such as Figure 9.19 put side by side.

However, for more than five variables the map method is not very convenient in use and it is preferable to use alternative methods. These will be discussed in the next chapter.

Simplification for NOR/NAND Circuits

In general, when the final logic circuit is to be drawn using NOR gates, the simplest conjunctive normal form is better than the simplest disjunctive normal form. This arises since a product of sums generates pairs of interior NOTs when converted into NOR gates and these are then redundant and may be cancelled. The disjunctive normal form also gives rise to a final negation at the end of the circuit which is frequently an additional logic level. Thus, to form a NOR circuit, the zeros of the Karnaugh map should be considered rather than the ones.

This procedure is reversed if a NAND circuit is required, since in this case the sum of products from the conjunctive normal form is advantageous. In both cases, however, loops giving non-negated literals are to be preferred to those giving negations (if a choice arises).

It is possible to use Karnaugh maps to provide other manipulations of Boolean functions than the simplest normal form. In a circuit consisting of NOR/NAND gates, a normal form may not lead to the minimum number of gates. If good use can be made of loops giving non-negated literals, the following method of map factorization may yield good results. The method is easier to comprehend in the formation of NAND circuits working basically with the 1s of the map. It can of course also be applied to NOR circuits by using duality.

Suppose that a NAND circuit is required for the Boolean function

$$f = (x + y + z)(x + \bar{y} + z)(\bar{x} + \bar{y} + \bar{z})$$

The Karnaugh map is shown in Figure 9.20. Since a NAND circuit is required,

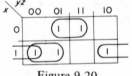

Figure 9.20

the simplest disjunctive normal form would be considered and the 1s of the map should be looped. The best grouping that can be done is to form three loops of two elements yielding

$$f = \bar{x}z + x\bar{y} + x\bar{z}$$

which gives a NAND circuit of seven gates after removing redundancies, as
shown in Figure 9.21.

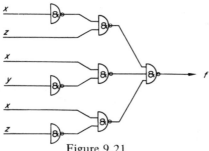

Figure 9.21

The simplest disjunctive normal form above can be further simplified
algebraically to give

$$f = \bar{x}z + x \cdot (\bar{y} + \bar{z})$$

This would take a circuit of five gates. If a conjunctive normal form is attempted,
the expression

$$f = (x + z)(\bar{x} + \bar{y} + \bar{z})$$

is obtained and a circuit would require seven gates.

Returning to the Karnaugh map of the function shows that it can nearly be
grouped into two loops of four elements as in Figure 9.22, except for one

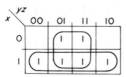

Figure 9.22

difficult zero in square 111 representing P_7. The function required than is
given by

$$x \cdot \bar{P}_7 + z \cdot \bar{P}_7$$

in which the unwanted term P_7 is inhibited from each of the two loops in which
it occurs. Thus

$$f = x \cdot (\overline{x\,y\,z}) + z \cdot (\overline{x\,y\,z})$$

The inhibited loop P_7 is obtained naturally with a NAND gate since

$$\overline{x\,y\,z} = x|y|z$$

The circuit for this form of f is shown in Figure 9.23 and requires only four
gates.

This type of manipulation depends very much on the skill of the designer in selecting suitable groupings of non-negated literals and will not always produce the minimum circuit. It does, however, indicate how a Karnaugh map can help in manipulating expressions into a desired form. If the negations of the variables

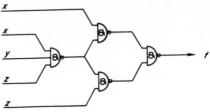

Figure 9.23

are already available from previous parts of a circuit, the simplest disjunctive normal form would also require only four NAND gates and would only be a two-level circuit.

Logic Operations with Karnaugh Maps

Since Karnaugh maps are simply a rearrangement of Venn diagrams which are representations of sets, it might be expected that the operations of the algebra

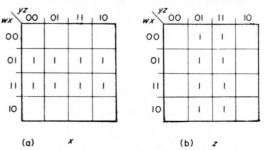

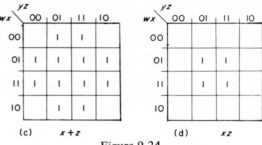

Figure 9.24

of classes might apply directly to the maps themselves. This is indeed the case if the universal set I is considered to be the complete Karnaugh map, while its subsets are all the $(2^{2^n} - 1)$ possible combinations of the individual squares. This means that logic operations may be applied directly to Karnaugh maps.

The map of Figure 9.24(a) represents the single variable x while that of Figure 9.24(b) represents z. The union of these two maps is shown in Figure 9.24(c) and can be obtained by forming the logical sum of each corresponding pair of squares of the original maps. In this operation any square not marked with a 1 is considered to be a 0 as usual. The map of Figure 9.24(c) therefore represents $x + z$.

In a similar manner to find the map representing the logical product $x\,z$, the intersection of the two maps is required and may be obtained as in Figure 9.24(d) by actually taking the logical product of corresponding pairs. Thus just the squares in which both of the original maps contain 1s contribute to the result.

These operations can be extended to any number of maps provided that all the maps represent functions of the same variables and any propositional or logical connective may be applied. Consider the example used earlier in the chapter with the function

$$f = (\overline{x} \to y)\ \&\ (y \not\equiv z)$$

This is carried out step by step in Figure 9.25 applying each operation directly to the individual squares of the maps. If the final diagram is compared with that of Figure 9.15, they are seen to be identical.

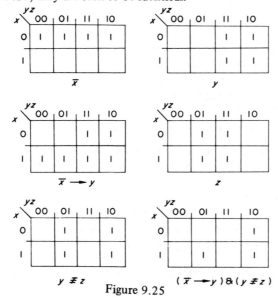

Figure 9.25

'Don't care' conditions can also be included in these operations if the following tables are used:

$$X . 1 = X, \qquad X + 1 = 1$$
$$X . 0 = 0, \qquad X + 0 = X$$

where X denotes the 'don't care' condition.

References and Further Reading

Karnaugh, M., 1953, 'The map method for synthesis of combinational logic circuits', *Trans. Am. Inst. Elect. Engrs.*, 72, Nov., 593–9.

Maley, G. A., and Earle, R., 1963, *The Logic Design of Transistor Digital Computers*, Prentice-Hall, Englewood Cliffs, N.J.

Veitch, E. W., 1952, 'A chart method for simplifying truth functions', *Proc. Conf. ACM*, May, pp. 127–33.

Exercises

1. Find the simplest disjunctive normal form using a Karnaugh map for the following functions of three variables:
   (a) $\bar{x}\bar{y}z + \bar{x}yz + x\bar{y}z + xyz$
   (b) $\bar{x}\bar{y}\bar{z} + x\bar{y}z + \bar{x}yz + xy\bar{z}$
   (c) $\bar{x} + x\bar{y}\bar{z} + x\bar{y}z$
   (d) $\bar{x}\bar{z} + \bar{y}z + x\bar{y} + \bar{x}y + \bar{x}z + \bar{y}\bar{z} + \bar{x}\bar{y}$

2. Determine the simplest disjunctive normal form of the following:
   (a) $\bar{w}\bar{x}yz + \bar{w}x\bar{y}\bar{z} + \bar{w}x\bar{y}z + \bar{w}xyz + w\bar{x}\bar{y}z + w\bar{x}y\bar{z} + wx\bar{y}z$
       $+ wxy\bar{z} + wxyz$
   (b) $\bar{w}\bar{y} + xy + \bar{y}z + w\bar{y}$
   (c) $\bar{w}\bar{x}\bar{y} + w\bar{x}\bar{y} + x\bar{y}z + wy\bar{z} + \bar{w}\bar{x}y\bar{z}$
   (d) $\bar{w}\bar{x}\bar{y}\bar{z} + \bar{w}\bar{x}\bar{y}z + \bar{w}\bar{x}yz + \bar{w}x\bar{y}z + \bar{w}xy\bar{z} + w\bar{x}\bar{y}\bar{z} + w\bar{x}\bar{y}z$
       $+ w\bar{x}y\bar{z} + wx\bar{y}\bar{z} + wxy\bar{z} + wxyz$

3. Find the simplest conjunctive normal form for each of the expressions of Questions 1 and 2.

4. Find both the simplest disjunctive and conjunctive normal forms for the following designation numbers which are expressed in terms of a standard basis:
   (a) 1111 1110
   (b) 1111 1110 1111 1110
   (c) 1111 1110 1111 1110 1111 1110 1111 1110
   (d) 1010 0101 1010 0101
   (e) 1111 1111 1100 1101

5. Find the simplest normal form for
   (a) $(y + \bar{z})(x + y + \bar{z})(\bar{x} + y + \bar{z})$
   (b) $w + x + y + z$
   (c) $(w + x + y + z)(w + x + y + \bar{z})(\bar{w} + x + y + z)(\bar{w} + x + y + \bar{z})$

6. Use the given 'don't care' conditions to find the simplest disjunctive normal form for the following:
   (a) $wz + yz + wx\bar{y}$
      with 'don't care' $\bar{w}\bar{y}$
   (b) $x + z$
      with 'don't care' $\bar{x}\bar{z}$
   (c) $(w + y)(\bar{w} + \bar{y})$
      with 'don't care' $\bar{w}y + w\bar{y}\bar{z}$

7. Simplify the functions of five variables:
   (a) $f(v, w, x, y, z) = \bar{v}x\bar{y} + v\bar{x}y + v\bar{w}z + vx$
   (b) $f(p, q, r, s, t) = P_0 + P_1 + P_9 + P_{16} + P_{17} + P_{24} + P_{25}$

8. Find all the simplest disjunctive normal forms of the following:
   (a) $f(w, x, y, z) = P_1 + P_2 + P_4 + P_5 + P_6 + P_{11} + P_{12} + P_{13} + P_{14} + P_{15}$
   (b) $f(p, q, r, s) = P_0 + P_2 + P_3 + P_7 + P_8 + P_9 + P_{11} + P_{13}$
   (c) $f(w, x, y, z) = S_0 . S_5 . S_{10} . S_{15}$

9. Find all the simplest conjunctive normal forms of the examples of Question 8. From both sets of results determine the best logic circuit consisting of AND, OR, NOT elements, taking your criterion of 'best' as the circuit with the least total number of inputs into the gates. (The normal forms should be further factored, where possible, before drawing the logic circuits.)

10. Design a circuit using NOR units for the Boolean function
    $f = (x \not\equiv y) . (y \not\equiv z)$, given that y and z are never both true together.

11. Find the simplest normal form that will lead to the best NOR circuit for the function

$$f(w, x, y, z) = \overline{w}\,\overline{x}\,y\,z + \overline{w}\,x\,\overline{y}\,z + \overline{w}\,x\,y\,\overline{z} + \overline{w}\,x\,y\,z + w\,\overline{x}\,\overline{y}\,\overline{z} + w\,\overline{x}\,y\,\overline{z} + w\,\overline{x}\,y\,z$$

What result would you get if $\overline{w}\,x\,\overline{y}\,z$ was a 'don't care' condition?

12. Determine the NAND circuit using the fewest number of gates and making use of the method of inhibiting unwanted terms where possible:

(a) $f(x, y, z) = \overline{x}\,\overline{y}\,z + \overline{x}\,y\,z + x\,\overline{y}\,\overline{z} + x\,y\,\overline{z}$
    with 'don't care' term $x\,\overline{y}\,z$

(b) $f(w, x, y, z) = P_1 + P_3 + P_4 + P_6 + P_7 + P_9 + P_{11} + P_{12} + P_{13} + P_{14}$

(c) $f(w, x, y, z) = P_0 + P_2 + P_3 + P_4 + P_6 + P_7 + P_8 + P_{10} + P_{11}$

Chapter 10
Simplification by Tabular Methods

Introduction

As was seen in the previous chapter, Karnaugh maps provide a very simple means of finding the simplest normal forms for Boolean functions of four or five variables or less. For more variables, this method becomes extremely unwieldy and various tabular methods have been developed. These are based mainly on a method by W. V. Quine and use the same criterion for minimality, viz. the disjunctive normal form with the fewest occurrences of literals. They aim at finding all the '*prime implicants*' of the function and then selecting from these a set to give the simplest disjunctive normal form.

Prime Implicants

Firstly, the definition of an *implicant*. An implicant p of a function f is a product of literals so that $p \rightarrow f$ is a tautology (or $p \Rightarrow f$). That is to say that the combination $p = 1, f = 0$ never occurs for any assignment of 0s and 1s to the variables of f. In terms of designation numbers, $\#p$ cannot have units in positions where $\#f$ has zeros. (This means that $\#p$ necessarily has a less or equal number of 1s than $\#f$.) If p and p^* are two products of literals such that $p \rightarrow p^*$ is a tautology, then p^* is formed from p by deleting some of the literals of p. For example

$$x\bar{y}z \Rightarrow x\bar{y}$$

A *prime* implicant of a function f is an implicant p that contains no shorter conjunction of literals that is also an implicant of f. That is, there does not exist $p*$ such that

$$p \Rightarrow p* \quad \text{and} \quad p* \Rightarrow f$$

Now suppose the simplest disjunctive normal form for a function f consists of the logical sum of a set of product terms p_i,

$$f = p_1 + p_2 + \ldots + p_n$$

For each of these p_i, $\# p_i$ cannot have a unit in a position where $\# f$ has a zero, otherwise the disjunction would contain a unit in that position and could not be equiveridic to f. Hence

$$p_i \Rightarrow f$$

and therefore *each p_i is an implicant of f.*

This expression is also required to be the simplest disjunctive normal form. No p_i therefore can be replaced by a different implicant p_i^* such that $p_i \Rightarrow p_i^*$ since then p_i^* would contain fewer literals than p_i and an even simpler form could be provided. *Therefore each p_i must be prime.*

The minimization of a function therefore consists of two stages.

(i)   Find *all* the prime implicants.

(ii) Select a subset of these to give the simplest disjunctive normal form.

Methods of Obtaining Prime Implicants

Four methods of finding all the prime implicants of a Boolean function will be demonstrated. The same function will be used as the example for all four methods, namely

$$f = P_1 \vee P_3 \vee P_4 \vee P_5 \vee P_6 \vee P_7 \vee P_8 \vee P_{12}$$
$$= \overline{w}\overline{x}\overline{y}z + \overline{w}\overline{x}yz + \overline{w}x\overline{y}\overline{z} + \overline{w}x\overline{y}z + \overline{w}xy\overline{z} + \overline{w}xyz + w\overline{x}\overline{y}\overline{z} + wx\overline{y}\overline{z}$$

(a) *Karnaugh map.* This will be used to illustrate the concept of prime implicants rather than a practical method of finding them. It is a very simple method of obtaining prime implicants but, as has already been mentioned, is only suitable for up to about four or five variables.

The prime implicants consist of *all* the largest possible groupings that can be obtained. Any legitimate grouping is an implicant of the function. Choosing largest possible groupings ensures that the implicants are prime. Since all the prime implicants are wanted at this stage, all possible largest groupings should be included. These are shown for the above function f in Figure 10.1.

From this Karnaugh map, the prime implicants of f are $w\bar{y}\bar{z}$, $x\bar{y}\bar{z}$, $\bar{w}x$, $\bar{w}z$. The function is equiveridic to the logical sum of these,

$$f = w\bar{y}\bar{z} + x\bar{y}\bar{z} + \bar{w}x + \bar{w}z$$

It can also be seen from the map that it is the term $x\bar{y}\bar{z}$ which is redundant in this case, since this is included in the disjunction of the remaining loops.

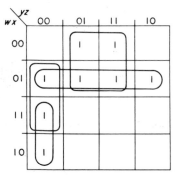

Figure 10.1 Prime implicants of function f

(b) *An algebraic method.* The *consensus* of two product terms that oppose each other in exactly one literal is the conjunction of all the other literals. That is to say that for precisely one of the variables (y, say), y appears in one term and $\bar{y}$ in the other. For example, the consensus of

$$wx\bar{y} + wy\bar{z}$$

is

$$wx\bar{z}$$

The consensus thus formed must be an implicant of the original disjunction.

A product p *includes* a product q if and only if all the literals of q are contained in the product p.

A disjunction of all the prime implicants of an expression f may be obtained by repeatedly taking the logical sum of the consensus of all products and dropping any including product terms until no further change can be made.

Applying this procedure to the same function f gives the following steps:

$$\bar{w}\bar{x}yz + \bar{w}\bar{x}yz + \bar{w}x\bar{y}\bar{z} + \bar{w}x\bar{y}z + \bar{w}xy\bar{z} + \bar{w}xyz + w\bar{x}\bar{y}\bar{z} + wx\bar{y}\bar{z}$$

(add consensus of all pairs of products)

$$+ \bar{w}\bar{x}z + \bar{w}\bar{y}z + \bar{w}yz$$
$$+ \bar{w}x\bar{y} + \bar{w}x\bar{z} + x\bar{y}\bar{z}$$
$$+ \bar{w}xz + \bar{w}xy + w\bar{y}\bar{z}$$

All including terms must now be dropped. This removes the whole of the original expression for f and leaves only the second part. The process is then repeated by again adding the consensus of all pairs of products. Thus to the above nine terms are added

$$+ \ \bar{w}\bar{y}z + \bar{w}z \quad + \bar{w}yz$$
$$+ \ \bar{w}z \quad + \bar{w}x\bar{y} + \bar{w}x\bar{y}$$
$$+ \ \bar{w}xz + \bar{w}xz + \bar{w}xy$$
$$+ \ \bar{w}x \quad + \bar{w}x \quad + \bar{w}x\bar{y}$$
$$+ \ \bar{w}x\bar{z} + \bar{w}x\bar{y} + \bar{w}x\bar{z}$$

Dropping all including terms gives

$$w\bar{y}\bar{z} + x\bar{y}\bar{z} + \bar{w}x + \bar{w}z$$

Attempting the procedure again, no new consensus terms can be formed. This final expression is therefore the disjunction of all the prime implicants.

(c) *McClusky's tabular method.* The method to be explained here is E. J. McClusky's improvement of W. V. Quine's original method. It will be written out in terms of designation numbers and is essentially based on the simplification

$$x\bar{y}\bar{z} + x\bar{y}z = x\bar{y}(\bar{z} + z)$$
$$= x\bar{y}$$

Take the standard basis $b[z, y, x]$

$$\#z = 0101 \quad 0101$$
$$\#y = 0011 \quad 0011$$
$$\#x = 0000 \quad 1111$$

$$\begin{array}{c} 0 \\ x\bar{y}\bar{z} \text{ is represented by column } 4 = 0 \\ 1 \end{array}$$

$$\begin{array}{c} 1 \\ x\bar{y}z \text{ is represented by column } 5 = 0 \\ 1 \end{array}$$

and it is seen that they differ in exactly one row. Two product terms that differ in exactly one row can always be simplified in this fashion. This corresponds to a loop of two adjacent squares on a Karnaugh map.

The designation number of the expression f to be simplified with respect to the basis $b[z, y, x, w]$ is

$$0101 \quad 1111 \quad 1000 \quad 1000$$

Write out all the columns of the basis that have units in the designation number..

1	3	4	5	6	7	8	12
1	1	0	1	0	1	0	0
0	1	0	0	1	1	0	0
0	0	1	1	1	1	0	1
0	0	0	0	0	0	1	1

Since the only terms that can be combined differ in exactly one row, rearrange the order of the columns according to the number of 1s in them.

```
No. of 1s: 0      1            2        3
           1 0 0  1 1 0 0      1
           0 0 0  1 0 1 0      1
           0 1 0  0 1 1 1      1
           0 0 1  0 0 0 1      0
           √ √ √  √ √ √ √      √
```

Comparisons are then made between adjacent sections. For example, every column in section 1 must be compared with every column in section 2. When a pair of columns differ in one row, they are both checked with a √. As they are combined, the resulting column is written out with the row in which they differ now marked with the composite symbol ♦.

```
1 1 ♦ 0 0 0    1 1 ♦
♦ 0 0 ♦ 0 0    1 ♦ 1
0 ♦ 1 1 1 ♦    ♦ 1 1
0 0 0 0 ♦ 1    0 0 0
√ √ √ √          √ √ √
```

The process is then repeated. This time the ♦ columns must also match. At this stage, loops of four on the Karnaugh map are being sought. This gives

```
1 1 ♦ ♦
♦ ♦ ♦ ♦
♦ ♦ 1 1
0 0 0 0
```

Two columns are duplicated so that one pair should be discarded. As there is only one section left, no further comparisons can be made and the process is terminated.

The prime implicants consist of the *unchecked* columns, which represent all loops which have not been able to combine with still larger loops. These are

$$0 \quad 0 \quad 1 \quad φ$$
$$0 \quad 0 \quad φ \quad φ$$
$$1 \quad φ \quad φ \quad 1$$
$$φ \quad 1 \quad 0 \quad 0$$

$$φ$$
$$φ$$

A column of the standard basis such as $\begin{matrix}1\\0\end{matrix}$ represents w equal to 0, x equal to 1, both values of y, and both values of z, i.e. the product term $\bar{w}x$ since y and z do not occur. It may be more convenient in writing to transpose the column into the row form of 0   1   φ   φ. The four prime implicants are

$$φ \quad 1 \quad 0 \quad 0 = x\bar{y}\bar{z}$$
$$1 \quad φ \quad 0 \quad 0 = w\bar{y}\bar{z}$$
$$0 \quad φ \quad φ \quad 1 = \bar{w}z$$
$$0 \quad 1 \quad φ \quad φ = \bar{w}x$$

(d) *Decimal method.* This is similar to the tabular method just described but is written out using the decimal notation as a convenient shorthand.

The decimal numbers of the columns are listed instead of the columns themselves but they must of course still be arranged according to the number of 1s in the column. It is also more convenient to lay out the numbers down the page and to carry out subsequent combinations across the page. This gives the lay-out for the worked example shown in Table 10.1.

Table 10.1. Decimal method of obtaining prime implicants

1 √	1 (2)√	1 (2, 4)
4 √	1 (4)√	
8 √	4 (1)√	4 (1, 2)
3 √	4 (2)√	
5 √	4 (8)	
6 √	8 (4)	
12 √	3 (4)√	
7 √	5 (2)√	
	6 (1)√	

The first column shows the original decimal numbers grouped into sections according to the numbers of 1s in the appropriate column of the designation number basis. (In this instance, the sections represent 1, 2, and 3 ones respectively.) The top section of column 1 must first be compared with the next section to determine if any of the decimal numbers *differ by a power of 2* (including $2^0 = 1$). If so, the first of the pair of numbers is written in column 2 followed by the actual difference in brackets. Thus the first correspondence here is the 1 from section 1 with the 3 from section 2, with a difference of $2^1 = 2$, giving the insertion 1 (2) at the top of column 2. As each successful correspondence is made, the entries are checked with $\checkmark$. Every number of section 1 is compared in this manner with every larger number in section 2. This gives the top six entries of column 2.

Next, the same process is continued but comparing the second section with the third section, and so on. This gives all the possible loops of 2 recorded in column 2.

Now the loops of 4 must be found by a similar comparison between the sections of column 2. This time, however, the number in brackets must also match before a successful correspondence is made. Comparing section 1 of column 2 with section 2, the first success is 1 (2) with 5 (2). These differ by $2^2 = 4$ and both have the same bracketed number 2. This is entered in a new column as 1 (2, 4) with both sets of differences retained in brackets, and the two constituents are checked. If the same loop appears more than once, it does not need to be entered again. In this case, comparing 1 (4) with 3 (4) gives a loop 1 (4, 2). This is identical with the loop 1 (2, 4) already entered with the combinations written in a different order. 1 (2, 4) represents the combinations of 1, 3, 5, 7 and 1 (4, 2) represents 1, 5, 3, 7.

The process is repeated until no further comparisons can be made. In every case, for a successful correspondence all the terms in brackets must be identical.

As before, the unchecked entries give the prime implicants. They are as follows:

$$4\,(8); \quad 8\,(4); \quad 1\,(2,4); \quad 4\,(1,2)$$

The term outside the bracket in each case represents the first column of the designation number that contains a 1. The other entries which represent the combined elements ◊ must be reconstituted by adding all possible combinations of the numbers in the brackets. For example 4 (1, 2, 8) if it occurred would represent the looping of the decimal numbers 4, 5, 6, 7, 12, 13, 14, 15, in the combined term ◊ 1 ◊ ◊ or x.

The prime implicants of the example are

$$
\begin{aligned}
4\,(8) \quad &= ◊ \ 1 \ 0 \ 0 = x\bar{y}\bar{z} \\
8\,(4) \quad &= 1 \ ◊ \ 0 \ 0 = w\bar{y}\bar{z} \\
1\,(2,4) &= 0 \ ◊ \ ◊ \ 1 = \bar{w}z \\
4\,(1,2) &= 0 \ 1 \ ◊ \ ◊ = \bar{w}x
\end{aligned}
$$

Selection of Prime Implicants

The disjunction of the prime implicants contains *all* the possible *largest groupings* of the variables and can often be longer than the original expression. It is therefore necessary to select, from all the prime implicants found, those that give the minimum number of literals in the final disjunctive normal form. Unfortunately there is no known systematic procedure for obtaining this simplest form directly. What must be done is to obtain all possible choices of subsets of the prime implicants that represent the given function without any redundancy. From all these possible representations, the one with the minimum number of literals can be selected or else the circuit designer can choose the one he prefers on other grounds.

The set of prime implicants is obtained from a prime implicant chart as shown in Table 10.2. All the columns of the designation number that contain units are shown along the top. The chart itself shows which columns are covered by each of the prime implicants. The entries on the chart can be made with any symbol such as 1 or x but it often saves mistakes if the actual column number is written.

Table 10.2. Prime implicant chart

Prime implicant		1	3	4	5	6	7	8	12
4 (8)	$x\bar{y}\bar{z}$			4					12
8 (4)	$w\bar{y}\bar{z}$							8	12
1 (2, 4)	$\bar{w}z$	1	3		5		7		
4 (1, 2)	$\bar{w}x$			4	5	6	7		

Note that

$$\bar{w}x = 0\ 1\ \phi\ \phi$$

which consists of columns

$$4,\quad 4+1=5,\quad 4+2=6,\quad 4+1+2=7$$

as in the decimal method for 4 (1, 2).

A selection of prime implicants must be made to cover all these entries. Some of the prime implicants must be chosen since they are the only ones that cover a particular position. For example, columns 1 and 3 are only covered by $\bar{w}z$. These are known as *essential* prime implicants and are marked first on the chart, giving Table 10.3.

The final step is to choose the non-essential prime implicants to cover all the remaining entries in the best possible way, giving fewest occurrences of literals. In this particular example, the essential prime implicants already cover the remaining entries of columns 4 and 12 so that no further terms need be chosen. The simplest disjunctive normal form is therefore

$$f = w\bar{y}\bar{z} + \bar{w}z + \bar{w}x$$

In normal working, the chart would of course only be drawn out once and the entries can be obtained direct from the final columns of the McClusky Table 10.3. Essential prime implicants

Prime implicant	1 3 4 5 6 7 8 12
$x\bar{y}z$	4 12
$w\bar{y}\bar{z}$	8 12
$\bar{w}z$	1 3 5 7
$\bar{w}x$	4 5 6 7

minimization. It is not necessary to write out all the prime implicants at length, and they are often referred to by letters A, B, C, etc. On larger charts it may be desirable to cross out a column completely once it has been included in a chosen prime implicant.

A Further Complete Example

The following example will be worked straight through using the decimal method. Find the simplest disjunctive normal form for the Boolean expression

$$f = P_0 + P_2 + P_3 + P_4 + P_6 + P_7 + P_8 + P_9 + P_{10} + P_{13} + P_{14} + P_{15}$$

The designation number is 1011 1011 1110 0111. Arranging the columns in order of numbers of 1s and performing the minimization gives Tables 10.4.

Table 10.4. Decimal minimization of example

0 ✓	0 (2) ✓	0 (2, 4) D
2 ✓	0 (4) ✓	0 (2, 8) E
4 ✓	0 (8) ✓	2 (1, 4) F
8 ✓	2 (1) ✓	2 (4, 8) G
3 ✓	2 (4) ✓	6 (1, 8) H
6 ✓	2 (8) ✓	
9 ✓	4 (2) ✓	
10 ✓	8 (1) A	
7 ✓	8 (2) ✓	
13 ✓	3 (4) ✓	
14 ✓	6 (1) ✓	
15 ✓	6 (8) ✓	
	9 (4) B	
	10 (4) ✓	
	7 (8) ✓	
	13 (2) C	
	14 (1) ✓	

There are eight prime implicants found and they are labelled from A to H. The corresponding chart is shown in Table 10.5. The essential prime implicants are

Table 10.5. Prime implicant chart for example

Prime implicant	0	2	3	4	6	7	8	9	10	13	14	15
A							8	9				
B								9		13		
C										13		15
D	0	2		4	6							
E	0	2					8		10			
F		2	3		6	7						
G		2			6				10		14	
H					6	7					14	15

easily seen to be D and F. One method of covering the remaining columns is shown on the chart and consists of A, C, and G. Thus a possible representation is

$$f = A + C + D + F + G$$
$$= 100◊ + 11◊1 + 0◊◊0 + 0◊1◊ +\ \ ◊◊10$$
$$= w\overline{x}\,\overline{y} + wxz + \overline{w}\,\overline{z} + \overline{w}y + y\overline{z}$$

This in fact is not the minimum expression that can be obtained. A method of finding all suitable representations is given below.

Algebraic Choice of Prime Implicants

In a minimization, any combination of the non-essential prime implicants that do in fact cover all the remaining entries of the chart when added to the essential prime implicants will yield a possible representation of the Boolean expression. However, usually a non-redundant set is required, that is a set from which it is not possible to remove any implicant and still retain the covering of all the elements. There will normally be several possible sets to choose from and the best combination to give fewest occurrences of literals can only be found by comparing all the possible combinations. Many of the combinations can be rejected immediately on an inspection of the prime implicant chart. In choosing a set of prime implicants it should be remembered that the implicants that cover the most columns require the fewest literals to express them.

It is possible to use an algebraic method to obtain all the sets of non-redundant prime implicants. Each prime implicant is referred to by the appropriate letter indicated in the chart as in Table 10.5. A new Boolean function g is formed that takes the truth value 1 when a complete covering exists. It may be formed from the chart by considering each column in turn and writing the condition for a cover of that column. For example, column 0 requires $D \vee E$, i.e. the disjunction of all entries in that column. The overall condition must be that the conjunction of the sums formed from each column must be true. Therefore, in this example,

$$g = (D + E) \cdot (D + E + F + G) \cdot F \cdot D \cdot (D + F + G + H) \cdot (F + H)$$
$$\cdot (A + E) \cdot (A + B) \cdot (E + G) \cdot (B + C) \cdot (G + H) \cdot (C + H)$$

This expression must be simplified by Boolean algebra and multiplied out to give a disjunctive normal form. Owing to the form of the expression, the simplification is very easy and relies on the two absorption laws

$$x + xy = x$$
$$x(x + y) = x$$

In the present example, the two single factors $F \cdot D$ allow the partial reduction

$$F \cdot D \cdot (D + E) \cdot (D + E + F + G) \cdot (D + F + G + H) \cdot (F + H) = F \cdot D$$

giving

$$g = F \cdot D \cdot (A + E) \cdot (A + B) \cdot (E + G) \cdot (B + C) \cdot (G + H) \cdot (C + H)$$

As terms are multiplied out, further reductions can be made, e.g.

$$(A + E) \cdot (A + B) = A \cdot A + A \cdot B + A \cdot E + B \cdot E = A + B \cdot E$$

Continuing this process gives

$$g = A \cdot C \cdot D \cdot F \cdot G + B \cdot D \cdot E \cdot F \cdot H + A \cdot B \cdot D \cdot F \cdot G \cdot H$$
$$+ A \cdot B \cdot D \cdot E \cdot F \cdot H + A \cdot C \cdot D \cdot E \cdot F \cdot H$$
$$+ B \cdot C \cdot D \cdot E \cdot F \cdot G$$

If any product term in this expression is true, then g will be true and a cover of the prime implicant chart has been achieved. Thus each of these product terms represents a possible choice of non-redundant prime implicants for f. Any individual product term states that each of the prime implicants mentioned must be present for the cover. In the term $A \cdot C \cdot D \cdot F \cdot G$, each of A, C, D, F, and G prime implicants are necessary for a cover of f, i.e.

$$f = A + C + D + F + G$$

This corresponds to the solution that was obtained directly from the chart. It is seen that every term contains the factors $D \cdot F$ corresponding to the two essential prime implicants.

A choice must now be made from the six possible sets of prime implicants obtained to find the set that yields the fewest number of literals. There are two points to consider: (a) the number of implicants in the product; (b) the number of literals needed to express each implicant. Thus two of the product terms contain five implicants while the other four contain six implicants. Also the prime implicants A, B, and C require three literals while D, E, F, G, H require only two. Therefore the term using fewest literals will be that corresponding to $B.D.E.F.H$, i.e.

$$f = B.D.E.F.H$$
$$= 1\lozenge01 + 0\lozenge\lozenge0 + \lozenge0\lozenge0 + 0\lozenge1\lozenge + 1\lozenge1\lozenge$$
$$= w\bar{y}z + \bar{w}\bar{z} + \bar{x}\bar{z} + \bar{w}y + xy$$

'Don't Care' Conditions

In the minimization by Karnaugh map, it was shown how easy it was to include 'don't care' conditions. These conditions will also arise in the minimization of functions involving more variables than four and it is therefore desirable to include 'don't cares' in the chart minimization. This extension is in fact very simple. The 'don't care' terms should be expressed as either columns of the standard basis or their decimal equivalents in the same manner as any other product term in the function to be minimized. They are then included in the chart and treated in an identical fashion with the normal terms in order to find the prime implicants. In this manner, they will be used as 1s to form the largest loops whenever possible.

However, when it comes to selecting a set of prime implicants by the prime implicant chart, only the decimal numbers corresponding to the actual function are displayed. The 'don't care' terms are not included since, as they may be omitted at will, no cover of them is required. Thus, to summarize, all 'don't care' terms should be included to obtain the prime implicants but they should all be omitted in the selection of a suitable minimum set.

The method will be illustrated by a simple example in three variables only. Simplify the expression

$$f = \bar{y}z + xz$$

subject to the 'don't care' condition $\bar{x}y$. The first step is to obtain either the special disjunctive normal form or else the designation number to determine the positions of the 1 terms.

$$f = \bar{x}\bar{y}z + x\bar{y}z + xyz$$
$$= P_1 + P_5 + P_7$$

The 'don't care' terms are $\bar{x}yz + \bar{x}y\bar{z}$ or $P_2 + P_3$. The decimal chart minimization is shown in Table 10.6. There are two prime implicants $01\lozenge$ and $\lozenge\lozenge1$. The prime implicant chart is given in Table 10.7, and only columns 1, 5, 7 need to be included, not 2 and 3. Thus prime implicant A does not cover any of the columns, and B covers all three of the columns included on the chart. The

Table 10.6. Decimal chart including 'don't care' terms

1 ✓	1 (2) ✓	1 (2, 4) B
2 ✓	1 (4) ✓	
$\overline{3 \checkmark}$	2 (1) A	
5 ✓	$\overline{3 (4) \checkmark}$	
$\overline{7 \checkmark}$	5 (2) ✓	

Table 10.7. Prime implicant chart with 'don't care' terms omitted

Prime implicant	1 5 7
A	
B	1 5 7

other column that would be covered by B is 3 but this does not appear. Clearly B is an essential prime implicant and in fact covers all columns. The simplest disjunctive normal form is therefore

$$f = ⬦⬦1$$
$$= z$$

Other Tabular Methods

For five variables or less, the Karnaugh map method of minimization is undoubtedly the most convenient. Many methods have been investigated for the problem of minimization of functions with, say, six or more variables. The chart methods presented here have all been based on McClusky's method and are quite easy to understand and carry through. The main disadvantage of the method is the necessity of two distinct stages of first finding all the prime implicants and then having to select the minimum version using another chart.

Many alternative methods for simplification have been investigated, attempting in particular to perform the minimization with a single chart. One of these is presented in Ledley and consists of finding the essential prime implicants directly. Another method known as the 'Cranfield' method is given in Flegg and works more directly with the function variables rather than their coded binary or decimal forms. This does have the advantage that, when alternatives have to be considered, the actual variables appear and hence possible fan-in or fan-out limitations can be considered. While these methods save a certain amount of actual writing, they are more difficult to learn and apply. Details for these methods may be found in books by the authors quoted above (Ledley, 1960; Flegg, 1964).

References and Further Reading

Flegg, H. G., 1964, *Boolean Algebra and its Applications,* Blackie, London.
Ledley, R. S., 1960, *Digital Computer and Control Engineering,* McGraw-Hill, New York.
McCluskey, E. J., 1956, 'Algebraic minimization and the design of two-terminal contact networks', *M.I.T. Thesis.*
—— 1956, 'Minimization of Boolean functions', *Bell Syst. Tech. J.,* 1417—44.
Quine, W. V. O., 1952, 'The problem of simplifying truth functions', *Am. Math. Monthly,* **59,** 521—31.
—— 1955, 'A way to simplify truth functions', *Am. Math. Monthly,* **62,** 627—31.
—— 1959, 'On cores and prime implicants of truth functions', *Am. Math. Monthly,* **66,** 755—60.

Exercises

1. Find all the prime implicants of the following functions by using a Karnaugh map:

   (a) $f(x, y, z) = \bar{x} + xz$

   (b) $f(x, y, z) = \bar{x}\bar{y}\bar{z} + \bar{x}\bar{y}z + \bar{x}y\bar{z} + x\bar{y}z + xy\bar{z} + xyz$

   (c) $f(w, x, y, z) = xz + \bar{w}\bar{x}yz + w\bar{x}\bar{y}z + wxy\bar{z}$

   (d) $f(w, x, y, z) = P_0 + P_1 + P_3 + P_4 + P_5 + P_6 + P_7 + P_8 + P_9 + P_{10}$
       $+ P_{11} + P_{12} + P_{13} + P_{14}$

   (e) $f(v, w, x, y, z) = 1011 \quad 1000 \quad 1101 \quad 1101 \quad 1000 \quad 1000 \quad 0111 \quad 110$

2. Check the results of Question 1 by using the algebraic method of the consensus.

3. Use a prime implicant chart to obtain the simplest disjunctive normal form for each of the functions of Question 1.

4. Obtain the simplest disjunctive normal forms for the previous functions by applying the algebraic method to select the prime implicants.

5. Find the prime implicants of the functions of Question 1 by using the McClusky chart method.

6. Find the prime implicants of the functions of Question 1 by using the decimal version of the chart method.

7. Find the simplest disjunctive normal forms for the following:
   (a) $v \not\equiv w \not\equiv x \not\equiv y \not\equiv z$
   (b) $(((v \to w) \to x) \to y) \to z$
   (c) $(((v \downarrow w) \downarrow x) \downarrow y) \downarrow z$

8. Determine all the possible choices of non-redundant selections of prime implicants that will represent the following functions. Use the decimal method to find all the prime implicants, followed by an algebraic selection to produce non-redundant covers.
   (a) 1111 1110 0111 1101
   (b) 1101 0110 1011 1110
   (c) 1111 1111 1111 1111 1111 1111 1111 1110

9. Use a tabular method to determine the simplest disjunctive normal form of the expression in five variables

$$f(v, w, x, y, z) = S_2 . S_3 . S_6 . S_7 . S_{14} . S_{15} . S_{30} . S_{31}$$

given the 'don't care' condition $\bar{v} x$.

10. Determine the simplest disjunctive normal function for each of the following designation numbers expressed in terms of standard bases, where the symbol X denotes a 'don't care' condition.
   (a) X1X0 1100 XXX0 01XX
   (b) X1X0 0111 X0X1 1101
   (c) 1111 XXXX 1111 1111 XXXX 0000 1111 0011

Chapter 11
Logical Matrices

Introduction

Two types of logical matrices will be considered here, switching matrices and Boolean matrices. In switching matrices the elements will be 0, 1, a logical element or a function of such elements. In Boolean matrices, on the other hand, only the truth values, represented by 0 and 1, appear.

Switching Matrices

Listing the connections of all the nodes, the four-terminal circuit of Figure 11.1 yields Table 11.1, where the logical switching function is that connecting node r directly to node s without passing through any other nodes. This, when set out in matrix form, is known as the *connection matrix.* In the case of switches, which are bi-directional, it is a symmetric matrix having the form

$$\begin{bmatrix} 1 & 0 & x & y \\ 0 & 1 & v & w \\ x & v & 1 & z \\ y & w & z & 1 \end{bmatrix}$$

i.e. with all the diagonal elements 1. This is due to the fact that a node can be considered to be feeding itself. Denoting the matrix by the letter $\mathbf{Z}$ and the elements of Z by z_{rs}, where r indicates the row and s the column in which z_{rs} is to be found, then z_{rs} will be the logical function of the network connecting node r to node s *without involving any other node.* This is why there is a 0 for z_{12} and, owing to symmetry, for z_{21}, since there is no path connecting nodes

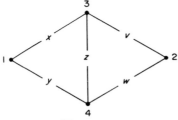

Figure 11.1

1 and 2 which is independent of any one of the other nodes. Hence, in Figure 11.1

$$z_{12} = z_{21} = 0$$
$$z_{13} = z_{31} = x$$
$$z_{14} = z_{41} = y$$
$$z_{23} = z_{32} = v$$
$$z_{24} = z_{42} = w$$
$$z_{34} = z_{43} = z$$
$$z_{11} = z_{22} = z_{33} = z_{44} = 1$$

These z_{rs} will be referred to as *connection functions* and a *transmission path* associated with two nodes will be the logical function resulting from the simplified Boolean sum of all the connection functions forming unbroken paths

Table 11.1

Transmitting node	Direct* logical function	Receiving node
1	0	2
1	x	3
1	y	4
2	v	3
2	w	4
3	z	4

* Not involving any other nodes.

between the two nodes. For example, the transmission path between nodes 1 and 3 is, in the case of Figure 11.1,

$$x + yz + ywv$$

The matrix of transmission paths will be referred to as a *terminal matrix* and thus the terminal matrix of Figure 11.1 is

$$
\begin{bmatrix}
1 & (xv + yw + xzw + yzv) & (x + yz + ywv) & (y + xz + xv \\
(vx + wy + wzx + vzy) & 1 & (v + wz + wyx) & (w + vz + vx \\
(x + zy + vwy) & (v + zw + xyw) & 1 & (z + vw + xy \\
(y + zx + wvx) & (w + zv + yxv) & (z + wv + yx) & 1
\end{bmatrix}
$$

Note that this matrix is symmetric.

Boolean Matrices

A Boolean matrix has as its elements only 0 or 1 representing *false* or *true* respectively. The simplest Boolean matrices have already been met in another guise, namely Karnaugh maps.

Figure 11.2

For example, the Karnaugh map of the three-variable function $\bar{x}y + \bar{y}(x \equiv z)$ is given by Figure 11.2 and its corresponding Boolean matrix by

$$
\begin{bmatrix}
1 & 0 & 1 & 1 \\
0 & 1 & 0 & 0
\end{bmatrix}
$$

As can be seen, from the above example, these matrices do not need to be square. Further, the logical operations AND, OR, and NOT apply to such matrices, element by element. For example, if

$$
X = \begin{bmatrix} 1 & 0 & 1 & 1 \\ 0 & 1 & 0 & 0 \end{bmatrix} \quad \text{and} \quad Y = \begin{bmatrix} 1 & 0 & 0 & 0 \\ 1 & 1 & 0 & 0 \end{bmatrix}
$$

then

$$
\bar{X} = \begin{bmatrix} 0 & 1 & 0 & 0 \\ 1 & 0 & 1 & 1 \end{bmatrix}
$$

$$
X + Y = \begin{bmatrix} 1 & 0 & 1 & 1 \\ 1 & 1 & 0 & 0 \end{bmatrix}
$$

and

$$
XY = \begin{bmatrix} 1 & 0 & 0 & 0 \\ 0 & 1 & 0 & 0 \end{bmatrix}
$$

Note that the latter two operations require that the matrices have the same dimensions. If this is not the case, then one of the matrices must be modified accordingly. For example, if it is desired to form **PQ**, where **P** is a 4 x 4 Boolean matrix based upon a four-variable map involving w, x, y, and z say,

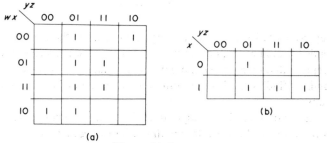

(a)

(b)

Figure 11.3

and **Q** is a 2 x 4 matrix based upon a three-variable map involving x, y, and z, then it is necessary to introduce w into **Q** *without changing the logical value of* **Q**. That is to say, the logical product of each of the terms of its disjunctive normal form with $(w + \overline{w})$ must be formed.

Let

$$P = \begin{bmatrix} 0 & 1 & 0 & 1 \\ 0 & 1 & 1 & 0 \\ 0 & 1 & 1 & 0 \\ 1 & 1 & 0 & 0 \end{bmatrix} \quad \text{and} \quad Q = \begin{bmatrix} 0 & 1 & 0 & 0 \\ 0 & 1 & 1 & 1 \end{bmatrix}$$

These are derived from the Karnaugh maps of Figure 11.3, which are themselves representations of

$$P = z(x + \overline{y}) + \overline{x}(w\overline{y} + \overline{w}y\overline{z})$$

and

$$Q = z(x + \overline{y}) + x(y + z)$$

Replacing the three-variable map by one involving w, as above, **Q** is obtained in the form

$$Q = \begin{bmatrix} 0 & 1 & 0 & 0 \\ 0 & 1 & 1 & 1 \\ 0 & 1 & 1 & 1 \\ 0 & 1 & 0 & 0 \end{bmatrix}$$

Thus

$$PQ = \begin{bmatrix} 0 & 1 & 0 & 0 \\ 0 & 1 & 1 & 0 \\ 0 & 1 & 1 & 0 \\ 0 & 1 & 0 & 0 \end{bmatrix}$$

i.e. PQ represents $z\,(\bar{y} + x)$. Hence the logical product of the two matrices corresponds to finding the Boolean function of the series connection of the two circuits represented by P and Q. Similarly, the logical sum of the two matrices corresponds to finding the Boolean function of the parallel connection of the two circuits.

The negation of a matrix corresponds to function inversion, as can be seen by referring to Figure 11.3 together with the following matrices:

$$Q = \begin{bmatrix} 0 & 1 & 0 & 0 \\ 0 & 1 & 1 & 1 \end{bmatrix}$$

and

$$\bar{Q} = \begin{bmatrix} 1 & 0 & 1 & 1 \\ 1 & 0 & 0 & 0 \end{bmatrix}$$

Two other logical operations, namely those of implication and equivalence (together with compound operations involving more than one logical connective) can also be carried out element by element on matrices having the same dimensions.

Let

$$A = \begin{bmatrix} 0 & 1 & 0 & 0 \\ 0 & 1 & 1 & 1 \end{bmatrix}, \qquad B = \begin{bmatrix} 1 & 0 & 1 & 0 \\ 0 & 0 & 0 & 1 \end{bmatrix}$$

and

$$C = \begin{bmatrix} 1 & 0 & 0 & 1 \\ 0 & 1 & 0 & 1 \end{bmatrix}$$

then

$$A \rightarrow B = \begin{bmatrix} 1 & 0 & 1 & 1 \\ 1 & 0 & 0 & 1 \end{bmatrix}$$

$$A \equiv B = \begin{bmatrix} 0 & 0 & 0 & 1 \\ 1 & 0 & 0 & 1 \end{bmatrix}$$

and

$$(A \not\equiv B)C = \begin{bmatrix} 1 & 0 & 0 & 0 \\ 0 & 1 & 0 & 0 \end{bmatrix}$$

If $P = \overline{w}\,y\,(\overline{x} + z) + x\,\overline{y}\,(w + \overline{z}) + x\,\overline{z}\,(w + \overline{y}) + w\,z\,(\overline{x} + \overline{y})$, i.e. P is a function of four variables w, x, y, and z, and $Q = \overline{z}(\overline{w} + y) + w\overline{y}z$, i.e. Q is a function of only three variables w, z, and y, find the matrix corresponding to $P \equiv Q$.

The matrix forms of the Karnaugh maps for P and Q are

$$P = \begin{bmatrix} 0 & 0 & 1 & 1 \\ 1 & 0 & 1 & 0 \\ 1 & 1 & 0 & 1 \\ 0 & 1 & 1 & 0 \end{bmatrix} \quad \text{and} \quad Q = \begin{bmatrix} 1 & 0 & 0 & 1 \\ 0 & 1 & 0 & 1 \end{bmatrix}$$

Hence, as in the previous example involving matrices of different sizes, the missing variable x must be introduced into the expression for Q giving rise to

$$Q = \overline{z}(\overline{w} + y)(x + \overline{x}) + w\overline{y}z(x + \overline{x})$$

This yields

$$Q = \begin{bmatrix} 1 & 0 & 0 & 1 \\ 1 & 0 & 0 & 1 \\ 0 & 1 & 0 & 1 \\ 0 & 1 & 0 & 1 \end{bmatrix}$$

and so

$$P \equiv Q = \begin{bmatrix} 0 & 1 & 0 & 1 \\ 1 & 1 & 0 & 0 \\ 0 & 1 & 1 & 1 \\ 1 & 1 & 0 & 0 \end{bmatrix}$$

As before, care must be taken to ensure that the Karnaugh maps, from which the matrix forms of P and Q are derived, employ the same lay-out of the variables w, x, y, and z.

Extension to Logical Matrices in General

The rules so far employed can be extended to logical matrices in general. If the three logical matrices W, Y, and Z have as their elements w_{rs}, y_{rs}, and z_{rs} respectively where these elements can be logical expressions, then

if	$W = Y$	then	$w_{rs} = y_{rs}$
	$W = \overline{Y}$		$w_{rs} = \overline{y}_{rs}$
	$W = YZ$		$w_{rs} = y_{rs}z_{rs}$
	$W = Y + Z$		$w_{rs} = y_{rs} + z_{rs}$
	$W = Y \rightarrow Z$		$w_{rs} = y_{rs} \rightarrow z_{rs}$
	$W = Y \equiv Z$		$w_{rs} = y_{rs} \equiv z_{rs}$

For example, if

$$\mathbf{Y} = \begin{bmatrix} (x+y) & 0 & (y+x\bar{z}) \\ 1 & (x \to y) & xyz \\ 0 & \bar{y} & \bar{z} \end{bmatrix}$$

and

$$\mathbf{Z} = \begin{bmatrix} (x+\bar{z}) & (y \equiv z) & x \\ yz & x & (x+\bar{z}) \\ x\bar{y}z & \bar{x} & yz \end{bmatrix}$$

then

$$\mathbf{YZ} = \begin{bmatrix} (x+\bar{z})(x+y) & 0 & x(y+x\bar{z}) \\ yz & x(x \to y) & (x+\bar{z})xyz \\ 0 & \bar{x}\bar{y} & 0 \end{bmatrix}$$

$$= \begin{bmatrix} (x+y\bar{z}) & 0 & x(y+\bar{z}) \\ yz & xy & xyz \\ 0 & \bar{x}y & 0 \end{bmatrix}$$

The Universal, Null, and Unit Matrices

Three special logical matrices are defined as follows. The *universal* matrix $\mathbf{U}$ is such that $u_{rs} = 1$ for all r and s. The *null* matrix $\mathbf{O}$ is such that $o_{rs} = 0$ for all r and s. Some useful identities involving $\mathbf{U}$ and $\mathbf{O}$ and any general logical matrix $\mathbf{X}$ are

$$\mathbf{O} + \mathbf{X} = \mathbf{X}$$
$$\mathbf{OX} = \mathbf{O}$$
$$\mathbf{U} + \mathbf{X} = \mathbf{U}$$
$$\mathbf{UX} = \mathbf{X}$$
$$\mathbf{O} + \mathbf{U} = \mathbf{U}$$
$$\mathbf{OU} = \mathbf{O}$$
$$\bar{\mathbf{O}} = \mathbf{U}$$
$$\bar{\mathbf{U}} = \mathbf{O}$$
$$\mathbf{O} \to \mathbf{X} = \mathbf{U}$$
$$\mathbf{U} \to \mathbf{X} = \mathbf{X}$$
$$\mathbf{O} \equiv \mathbf{X} = \mathbf{X}$$
$$\mathbf{U} \equiv \mathbf{X} = \mathbf{X}$$
$$\mathbf{X} \to \mathbf{X} = \mathbf{U}$$

In addition, $\mathbf{X} \to \mathbf{Y} = \mathbf{U}$ if $\mathbf{X} = \mathbf{Y}$ or, less strongly, (a) $\mathbf{X} + \mathbf{Y} = \mathbf{Y}$ or (b) $\mathbf{XY} = \mathbf{X}$. In particular, if $\mathbf{X} \to \mathbf{Y} = \mathbf{U}$ *and* $\mathbf{Y} \to \mathbf{X} = \mathbf{U}$, then $\mathbf{X} = \mathbf{Y}$. Also, if $\mathbf{X} \to \mathbf{Y} = \mathbf{U}$ *and*

$Y \rightarrow Z = U$, then $X \rightarrow Z = U$. The third special logical matrix is the *unit* matrix I such that $i_{rs} = 1$ for all $r = s$ otherwise $i_{rs} = 0$. This latter matrix is only of value when the logical matrix product operator defined below is used.

The Logical Matrix Product

To form a matrix product, as distinct from the conjunction previously discussed, a new operator denoted by the multiplication symbol x is defined as follows.

If Z is to be the matrix product of X with Y, then symbolically

$$Z = X \times Y$$

where

$$z_{rs} = \sum_{k=1}^{m} x_{rk} y_{ks}$$

$$= x_{r1} y_{1s} + x_{r2} y_{2s} + \ldots + x_{rm} y_{ms}$$

Note that the capital sigma is used here in the sense of *Boolean summation*, i.e. the disjunction of the terms $x_{rk} y_{ks}$ is formed.

The very nature of the logical matrix product demands that $X \times Y$ only has meaning when the number of columns of X is the same as the number of rows of Y. Further, if $X \times Y$ can be formed it does not follow that it will be the same as $Y \times X$, in fact the latter logical matrix product need not even exist.

As a first example, let

$$X = \begin{bmatrix} 1 & 0 & 1 & 0 \\ 1 & 1 & 0 & 0 \\ 1 & 0 & 0 & 1 \end{bmatrix} \quad \text{and} \quad Y = \begin{bmatrix} 0 & 1 \\ 1 & 1 \\ 0 & 0 \\ 1 & 1 \end{bmatrix}$$

then

$$X \times Y = \begin{bmatrix} 1 & 0 & 1 & 0 \\ 1 & 1 & 0 & 0 \\ 1 & 0 & 0 & 1 \end{bmatrix} \times \begin{bmatrix} 0 & 1 \\ 1 & 1 \\ 0 & 0 \\ 1 & 1 \end{bmatrix}$$

$$= \begin{bmatrix} 0 & 1 \\ 1 & 1 \\ 1 & 1 \end{bmatrix}$$

However, the product $\mathbf{Y} \times \mathbf{X}$ cannot be formed. Secondly, if

$$\mathbf{X} = \begin{bmatrix} 0 & x & \bar{y} & (x+y) \\ \bar{x} & 1 & 0 & xy \\ (\bar{x}+y) & y & (\bar{x}+\bar{y}) & \bar{y} \end{bmatrix}$$

and

$$\mathbf{Y} = \begin{bmatrix} 0 & 0 & 1 \\ 1 & 0 & 0 \\ 0 & 1 & 0 \\ 1 & 0 & 1 \end{bmatrix}$$

then

$$\mathbf{X} \times \mathbf{Y} = \begin{bmatrix} (x+(x+y)) & \bar{y} & (x+y) \\ (1+xy) & 0 & (\bar{x}+xy) \\ (y+\bar{y}) & (\bar{x}+\bar{y}) & ((\bar{x}+y)+\bar{y}) \end{bmatrix}$$

$$= \begin{bmatrix} (x+y) & \bar{y} & (x+y) \\ 1 & 0 & (\bar{x}+y) \\ 1 & (\bar{x}+\bar{y}) & 1 \end{bmatrix}$$

and

$$\mathbf{Y} \times \mathbf{X} = \begin{bmatrix} (\bar{x}+y) & y & (\bar{x}+\bar{y}) & \bar{y} \\ 0 & x & \bar{y} & (x+y) \\ \bar{x} & 1 & 0 & xy \\ (\bar{x}+y) & (x+y) & (\bar{x}+\bar{y}) & 1 \end{bmatrix}$$

Bearing in mind the restrictions imposed by the definition of the logical matrix product upon the number of rows and columns of the matrices involved, the list of useful identities can be extended as follows:

$$\mathbf{I} \times \mathbf{I} = \mathbf{I}$$
$$\mathbf{U} \times \mathbf{U} = \mathbf{U}$$
$$\mathbf{O} \times \mathbf{O} = \mathbf{O}$$
$$\mathbf{O} \times \mathbf{U} = \mathbf{O}$$
$$\mathbf{U} \times \mathbf{I} = \mathbf{U}$$
$$\mathbf{I} \times \mathbf{O} = \mathbf{O}$$
$$\mathbf{O} \times \mathbf{Z} = \mathbf{Z} \times \mathbf{O} = \mathbf{O}$$
$$\mathbf{I} \times \mathbf{Z} = \mathbf{Z} \times \mathbf{I} = \mathbf{Z}$$
$$\mathbf{X} \times (\mathbf{YZ}) = (\mathbf{X} \times \mathbf{Y})(\mathbf{X} \times \mathbf{Z})$$
$$\mathbf{X} \times (\mathbf{Y} + \mathbf{Z}) = (\mathbf{X} \times \mathbf{Y}) + (\mathbf{X} \times \mathbf{Z})$$
$$(\mathbf{XY}) \times \mathbf{Z} = (\mathbf{X} \times \mathbf{Z})(\mathbf{Y} \times \mathbf{Z})$$
$$(\mathbf{X} + \mathbf{Y}) \times \mathbf{Z} = (\mathbf{X} \times \mathbf{Z}) + (\mathbf{Y} \times \mathbf{Z})$$
$$\mathbf{X} \times (\mathbf{Y} \times \mathbf{Z}) = (\mathbf{X} \times \mathbf{Y}) \times \mathbf{Z}$$

In addition, if $Y \to Z = U$, then $(Y \times X) \to (Z \times X) = U$, for some X having the same dimensions as Y and Z but being otherwise arbitrary, and $(X \times Y) \to (X \times Z) = U$.

The Logical Matrix Transpose

The logical matrix transpose of Z is denoted by Z^T and is obtained from Z by interchanging the rows and columns, i.e. if $Z = z_{rs}$ then $Z^T = z_{sr}$. Hence, for switching matrices, if $Z^T Z = O$ then Z is skew-symmetric, whilst if $Z^T Z = I$ it is symmetric.

Input–Output Relationships

Given a connection matrix of a circuit, it is required to find the Boolean functions linking any input with any output. It is sufficient to eliminate all the nodes which are neither inputs nor outputs. It has been shown previously that the transmission path between nodes 1 and 3 of Figure 11.1 reduced to $x + yz + vwy$, so that if node 1 is considered to be the input node and node 3 the output, nodes 2 and 4 being internal, then the removal of nodes 2 and 4 would result in Figure 11.4.

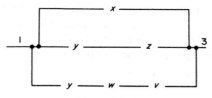

Figure 11.4

In this simple example, the method of reduction is shown in Figures 11.5 to 11.7 inclusive. Figure 11.5 is merely the bridge circuit of Figure 11.1 re-drawn, in

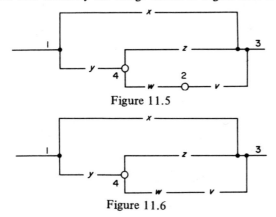

Figure 11.5

Figure 11.6

series-parallel form, to simplify the reduction. In Figure 11.6, node 2 has been eliminated and, in Figure 11.7, node 4. The final result can be left as in

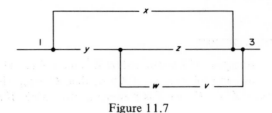

Figure 11.7

Figure 11.7, namely $x + y(z + wv)$ or it can be given in the form of Figure 11.4, namely $x + yz + vwy$.

Systematic Reduction of Switching Circuits

To reduce a switching circuit, from one consisting of n nodes to one consisting of $n - 1$ nodes, in a systematic manner the following technique can be used. If the nodes are numbered $1, 2, 3, \ldots, k, \ldots, n$ and it is desired to remove the node k, then replace all the z_{rs}, where $r \neq s$, by the expression $z_{rs} + z_{rk}z_{ks}$. For example, the circuit of Figure 11.8(a), in which node 3 is assumed to be an internal node, is equivalent to that of Figure 11.8(b). As far as the nodes 1 and 2 are concerned, z_{12} has been replaced, $z_{12} + z_{13}z_{32}$.

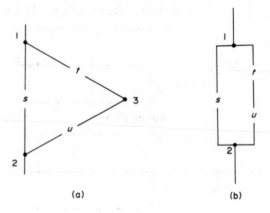

(a)   (b)

Figure 11.8

As a second example, consider the four-terminal network of Figure 11.9, with input at node 1 and output at node 2, nodes 3 and 4 being internal nodes.

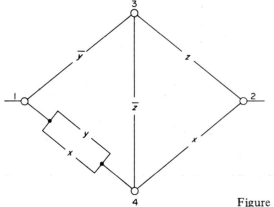

Figure 11.9

The full connection matrix for this network is

$$\begin{bmatrix} 1 & 0 & \bar{y} & (x+y) \\ 0 & 1 & z & x \\ \bar{y} & z & 1 & \bar{z} \\ (x+y) & x & \bar{z} & 1 \end{bmatrix}$$

and, eliminating node 4 by the technique just described, Figure 11.10 results with its 3 × 3 connection matrix

$$\begin{bmatrix} 1 & (x+y)x & (\bar{y}+\bar{z}(x+y)) \\ x(x+y) & 1 & (z+x\bar{z}) \\ (\bar{y}+\bar{z}(x+y)) & (z+\bar{z}x) & 1 \end{bmatrix}$$

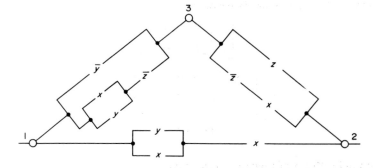

Figure 11.10

This simplifies, in the usual fashion, to

$$\begin{bmatrix} 1 & x & (\bar{z}+\bar{y}) \\ x & 1 & (x+z) \\ (\bar{z}+\bar{y}) & (x+z) & 1 \end{bmatrix}$$

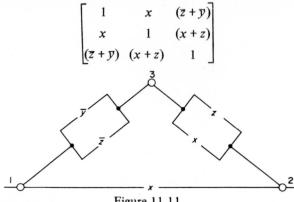

Figure 11.11

with the corresponding circuit of Figure 11.11. If node 3 is now eliminated the 2 x 2 connection matrix

$$\begin{bmatrix} 1 & x+(y+z)(x+z) \\ x+(\bar{y}+\bar{z})(x+z) & 1 \end{bmatrix}$$

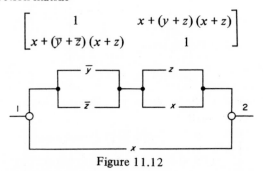

Figure 11.12

results with the corresponding circuit of Figure 11.12. This figure reduces further to Figure 11.13 and the matrix becomes

$$\begin{bmatrix} 1 & x+z\bar{y} \\ x+z\bar{y} & 1 \end{bmatrix}$$

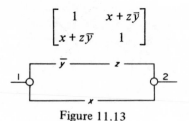

Figure 11.13

Reduction by Means of Logical Determinants

Another systematic approach, which does not, however, involve the manipulation of diagrams, makes use of determinants of logical matrices. The determinant of a

logical matrix Z is written as either det Z or $|Z|$, and this latter symbolism will
be used. For example, if

$$Z = \begin{bmatrix} z_{11} & z_{12} & z_{13} \\ z_{21} & z_{22} & z_{23} \\ z_{31} & z_{32} & z_{33} \end{bmatrix}$$

then $|Z|$ will be written as

$$\begin{vmatrix} z_{11} & z_{12} & z_{13} \\ z_{21} & z_{22} & z_{23} \\ z_{31} & z_{32} & z_{33} \end{vmatrix}$$

whenever it is necessary to display the elements of the determinant in full. Note
that a condition for a logical matrix to have a determinant is that the matrix
be square. Note also that the determinant of a logical matrix is either 0, 1, a
logical element or a logical expression, as can be seen from the following
definitions.

The Evaluation of Logical Determinants

The definition of the value of a determinant of order 1, i.e. one based upon a
matrix consisting of a single element, is simply the element itself. Hence, if

$$Z = [z_{11}] \qquad \text{then} \qquad |Z| = z_{11}$$

A determinant of order 2, i.e. one based upon the 2 x 2 matrix

$$Z = \begin{bmatrix} z_{11} & z_{12} \\ z_{21} & z_{22} \end{bmatrix}$$

is evaluated in two stages. The first stage entails expressing the determinant of
order 2 in terms of the Boolean sum of two products, each consisting of one
element of the original determinant and a determinant of order 1, i.e. one
degree less than the original determinant. In this case

$$|Z| = z_{11}|z_{22}| + z_{12}|z_{21}|$$

The second stage requires that the determinants of order 1 be replaced by their
logical values. Hence, since

$$|z_{22}| = z_{22} \qquad \text{and} \qquad |z_{21}| = z_{21}$$
$$|Z| = z_{11}z_{22} + z_{12}z_{21}$$

Note that the determinant of a 2 x 2 logical matrix reduces to the sum of two
terms, each of which consists of the product of two elements of the matrix.

This technique, of replacing a determinant of order m by m products of elements from the first row of the determinant with determinants of order $m-1$, will always hold and it can be shown, by induction, that a determinant of order m will reduce to the sum of $m!$ terms, each of which is a product of m elements of the original determinant. Further, the determinant of order $m-1$ which is to be associated with the element z_{1r} is obtained from the original determinant by deleting row 1 and column r.

For example, evaluate the determinant of the matrix $\mathbf{Z}$ where

$$\mathbf{Z} = \begin{bmatrix} z_{11} & z_{12} & z_{13} \\ z_{21} & z_{22} & z_{23} \\ z_{31} & z_{32} & z_{33} \end{bmatrix}$$

$$|\mathbf{Z}| = \begin{vmatrix} z_{11} & z_{12} & z_{13} \\ z_{21} & z_{22} & z_{23} \\ z_{31} & z_{32} & z_{33} \end{vmatrix}$$

$$= z_{11} \begin{vmatrix} z_{22} & z_{23} \\ z_{32} & z_{33} \end{vmatrix} + z_{12} \begin{vmatrix} z_{21} & z_{23} \\ z_{31} & z_{33} \end{vmatrix} + z_{13} \begin{vmatrix} z_{21} & z_{22} \\ z_{31} & z_{32} \end{vmatrix}$$

and since

$$\begin{vmatrix} z_{22} & z_{23} \\ z_{32} & z_{33} \end{vmatrix} = z_{22}z_{33} + z_{23}z_{32}$$

$$\begin{vmatrix} z_{21} & z_{23} \\ z_{31} & z_{33} \end{vmatrix} = z_{21}z_{33} + z_{23}z_{31}$$

and

$$\begin{vmatrix} z_{21} & z_{22} \\ z_{31} & z_{32} \end{vmatrix} = z_{21}z_{32} + z_{22}z_{31}$$

it follows that

$$|\mathbf{Z}| = z_{11}z_{22}z_{33} + z_{11}z_{23}z_{32} + z_{12}z_{21}z_{33} + z_{12}z_{23}z_{31} + z_{13}z_{21}z_{32} + z_{13}z_{22}z_{31}$$

Since the value of a logical determinant is unaltered either by interchanging any two rows or columns or by the operation of transposition, it is only necessary to consider the expansion of a determinant in terms of the first row as above. It may, however, be desirable to transpose and/or interchange rows or columns to reduce the amount of calculation required.

Example. Evaluate the determinant

$$\begin{vmatrix} (x+y) & \bar{x} & 0 & y \\ 1 & 0 & 0 & (x+y) \\ y & (\bar{x}+\bar{y}) & \bar{x}y & (x \equiv y) \\ 1 & 1 & 0 & 1 \end{vmatrix}$$

Since there exists one column in which all the elements except one are 0, i.e. false, the determinant is manipulated to make all the elements of the first row, except one, false. This is achieved in two steps. Firstly, the first and third columns are interchanged giving

$$\begin{vmatrix} 0 & \bar{x} & (x+y) & y \\ 0 & 0 & 1 & (x+y) \\ \bar{x}y & (\bar{x}+\bar{y}) & y & (x\equiv y) \\ 0 & 1 & 1 & 1 \end{vmatrix}$$

In the second step, transposition is used to achieve the desired result

$$\begin{vmatrix} 0 & 0 & \bar{x}y & 0 \\ \bar{x} & 0 & (\bar{x}+\bar{y}) & 1 \\ (x+y) & 1 & y & 1 \\ y & (x+y) & (x\equiv y) & 1 \end{vmatrix}$$

The evaluation now proceeds as described above, i.e. by expanding in terms of the first row. Thus

$$\mathbf{Z} = \bar{x}y \begin{vmatrix} \bar{x} & 0 & 1 \\ (x+y) & 1 & 1 \\ y & (x+y) & 1 \end{vmatrix}$$

$$= \bar{x}y \left\{ \bar{x} \begin{vmatrix} 1 & 1 \\ (x+y) & 1 \end{vmatrix} + 1 \begin{vmatrix} (x+y) & 1 \\ y & (x+y) \end{vmatrix} \right\}$$

$$= \bar{x}y(\bar{x}(1+(x+y)) + 1((x+y)+y))$$

$$= \bar{x}y(\bar{x}+x+y)$$

$$= \bar{x}y$$

Note that, if any row or column of a square matrix is null, i.e. it consists only of 0s, then the determinant of that matrix must be 0.

The transmission function for nodes r and s of a network can be found by evaluating the determinant of the connection matrix *with row r and column s deleted*. This means that, if the connection matrix is an $n \times n$ matrix, then the determinant to be evaluated will be of order $n-1$.

The transmission path for nodes 1 and 2 of Figure 11.9 has been determined in a previous example using a diagrammatic approach. Using determinants, the same result is obtained as follows. The connection matrix is

$$
\begin{bmatrix}
1 & 0 & \bar{y} & (x+y) \\
0 & 1 & z & x \\
\bar{y} & z & 1 & \bar{z} \\
(x+y) & x & \bar{z} & 1
\end{bmatrix}
$$

and the determinant of the matrix **A** which remains when row 1 and column 2 are removed is

$$
\begin{vmatrix}
0 & z & x \\
\bar{y} & 1 & \bar{z} \\
(x+y) & \bar{z} & 1
\end{vmatrix}
$$

Evaluating this yields the transmission function.

$$
\mathbf{A} = z \begin{vmatrix} \bar{y} & \bar{z} \\ (x+y) & 1 \end{vmatrix} + x \begin{vmatrix} \bar{y} & 1 \\ (x+y) & z \end{vmatrix}
$$

$$
= z(\bar{y} + \bar{z}(x+y)) + x(\bar{y}z + (x+y))
$$

$$
= z\bar{y} + \bar{z}(x+y) + x\bar{y}\bar{z} + x(x+y)
$$

$$
= x + z\bar{y}
$$

using the usual simplification techniques.

Input–Output Matrices

The matrix consisting of all transmission paths for every pair of input–output nodes *and no others* is termed an *input–output matrix* and is a special case of the terminal matrix defined earlier. The technique of producing an input–output matrix from the simpler connection matrix is as follows. Start by interchanging rows *and* columns so that all rows and columns referring to the non-terminal nodes are at the right-hand side and bottom of the matrix. These rows and columns are then eliminated one at a time using determinants.

For example, consider the switching circuit of Figure 11.14 with nodes 2, 3, and 5 as the input–output nodes.

The connection matrix is

$$
\begin{bmatrix}
1 & 0 & \bar{x} & \bar{y} & z \\
0 & 1 & y & z & x \\
\bar{x} & y & 1 & 0 & 0 \\
\bar{y} & \bar{z} & 0 & 1 & 0 \\
z & x & 0 & 0 & 1
\end{bmatrix}
$$

Interchanging rows 1 and 5 and columns 1 and 5 results in the matrix

$$\begin{bmatrix} 1 & x & 0 & 0 & z \\ x & 1 & y & z & 0 \\ 0 & y & 1 & 0 & \bar{x} \\ 0 & \bar{z} & 0 & 1 & \bar{y} \\ z & 0 & \bar{x} & \bar{y} & 1 \end{bmatrix}$$

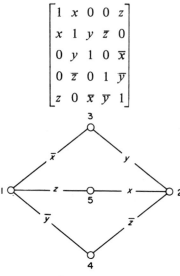

Figure 11.14

The fourth and fifth columns and rows are then removed using determinants. To remove the fifth column and row replace the original 5 x 5 matrix by the following 4 x 4 matrix,

$$\begin{bmatrix} 1 & \begin{vmatrix} x & z \\ 0 & 1 \end{vmatrix} & \begin{vmatrix} 0 & z \\ \bar{x} & 1 \end{vmatrix} & \begin{vmatrix} 0 & z \\ \bar{y} & 1 \end{vmatrix} \\ \begin{vmatrix} x & z \\ 0 & 1 \end{vmatrix} & 1 & \begin{vmatrix} y & 0 \\ \bar{x} & 1 \end{vmatrix} & \begin{vmatrix} z & 0 \\ \bar{y} & 1 \end{vmatrix} \\ \begin{vmatrix} 0 & z \\ \bar{x} & 1 \end{vmatrix} & \begin{vmatrix} y & 0 \\ \bar{x} & 1 \end{vmatrix} & 1 & \begin{vmatrix} 0 & \bar{x} \\ \bar{y} & 1 \end{vmatrix} \\ \begin{vmatrix} 0 & z \\ \bar{y} & 1 \end{vmatrix} & \begin{vmatrix} \bar{z} & 0 \\ \bar{y} & 1 \end{vmatrix} & \begin{vmatrix} 0 & \bar{x} \\ \bar{y} & 1 \end{vmatrix} & 1 \end{bmatrix}$$

Note that this matrix is also symmetric; the diagonal elements remaining 1, it will therefore only be necessary to discuss the construction of the elements above the diagonal. These elements are seen to be determinants of order 2 and the determinant which occurs in the position corresponding to the rth row and sth column is constructed as follows ($r < s < 5$). The element from the rth row and sth column of the original 5 x 5 matrix is placed in the top left position. The element from the rth row and 5th column is placed in the top right position and the element from the 5th row and sth column in the bottom left position.

The bottom right position is a 1. The determinants are evaluated and simplified
where necessary yielding

$$
\begin{bmatrix}
1 & x & z\bar{x} & z\bar{y} \\
x & 1 & y & \bar{z} \\
z\bar{x} & y & 1 & \bar{x}\bar{y} \\
z\bar{y} & z & \bar{x}\bar{y} & 1
\end{bmatrix}
$$

Repeating the procedure reduces the 4 x 4 matrix to

$$
\begin{bmatrix}
1 & \begin{vmatrix} x & z\bar{y} \\ \bar{z} & 1 \end{vmatrix} & \begin{vmatrix} z\bar{x} & z\bar{y} \\ \bar{x}\bar{y} & 1 \end{vmatrix} \\
\begin{vmatrix} x & zy \\ z & 1 \end{vmatrix} & 1 & \begin{vmatrix} y & \bar{z} \\ \bar{x}\bar{y} & 1 \end{vmatrix} \\
\begin{vmatrix} z\bar{x} & z\bar{y} \\ \bar{x}\bar{y} & 1 \end{vmatrix} & \begin{vmatrix} y & \bar{z} \\ \bar{x}\bar{y} & 1 \end{vmatrix} & 1
\end{bmatrix}
$$

which simplifies to

$$
\begin{bmatrix}
1 & x & z\bar{x} \\
x & 1 & (y + \bar{x}\bar{z}) \\
z\bar{x} & (y + \bar{x}\bar{z}) & 1
\end{bmatrix}
$$

and is the required input—output matrix. Note that the element x refers to
nodes 2 and 5, the conjunction $z\bar{x}$ refers to nodes 3 and 5 whilst the term
$(y + \bar{x}\bar{z})$ refers to nodes 2 and 3. In all cases, involving the interchange of
rows and columns, it is advisable to label the rows with their node numbers—
the numbering of the columns will be the same owing to the fact that the
symmetry is preserved. In the example shown the matrices would be labelled
as follows:

$$
\begin{array}{c}
1 \\ 2 \\ 3 \\ 4 \\ 5
\end{array}
\begin{bmatrix}
1 & 0 & \bar{x} & \bar{y} & z \\
0 & 1 & y & \bar{z} & x \\
\bar{x} & y & 1 & 0 & 0 \\
\bar{y} & \bar{z} & 0 & 1 & 0 \\
z & x & 0 & 0 & 1
\end{bmatrix}
$$

$$
\begin{array}{c}
5 \\ 2 \\ 3 \\ 4 \\ 1
\end{array}
\begin{bmatrix}
1 & x & 0 & 0 & z \\
x & 1 & y & \bar{z} & 0 \\
0 & y & 1 & 0 & \bar{x} \\
0 & \bar{z} & 0 & 1 & \bar{y} \\
z & 0 & \bar{x} & \bar{y} & 1
\end{bmatrix}
$$

$$
\begin{array}{c}
5 \\
2 \\
3 \\
4
\end{array}
\left[
\begin{array}{cccc}
1 & x & z\bar{x} & z\bar{y} \\
x & 1 & y & \bar{z} \\
z\bar{x} & y & 1 & \bar{x}\bar{y} \\
z\bar{y} & \bar{z} & \bar{x}\bar{y} & 1
\end{array}
\right]
$$

$$
\begin{array}{c}
5 \\
2 \\
3
\end{array}
\left[
\begin{array}{ccc}
1 & x & z\bar{x} \\
x & 1 & (y + \bar{x}\bar{z}) \\
z\bar{x} & (y + \bar{x}\bar{z}) & 1
\end{array}
\right]
$$

The circuit corresponding to this input–output matrix is shown in Figure 11.15.

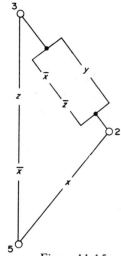

Figure 11.15

Whilst the above technique is universally applicable, in most simple cases two special procedures known as the star–delta and delta–star transformations

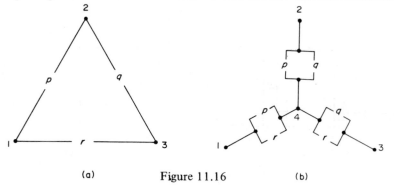

(a)   Figure 11.16   (b)

are all that are necessary to simplify the circuit. These are illustrated by the equivalent circuits of Figures 11.16 and 11.17. Note that, in forming Figure 11.16(b) from 11.16(a), an additional mode has been introduced and, in forming Figure 11.17(b) from 11.17(a), a node has been eliminated. As an

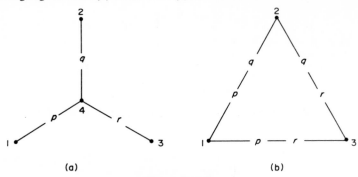

(a)   (b)

Figure 11.17

example, consider the reduction of Figure 11.18(a) to that of Figure 11.18(d). Figure 11.18(b) is the result of applying the delta–star transformation to nodes 2, 3, and 4 of Figure 11.18(a). Node 5 is the additional node introduced.

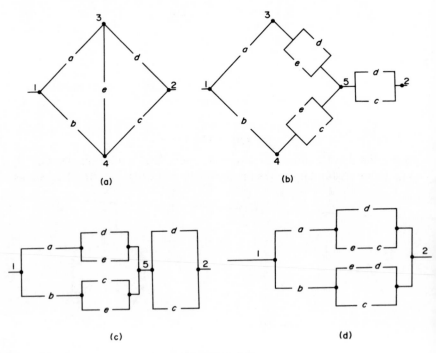

(a)   (b)

(c)   (d)

Figure 11.18

Figure 11.18(c) is merely Figure 11.18(b) re-drawn to make the series-parallel nature of the circuit more obvious. Finally, since

$$(y + x)(y + z) = y + xz$$

and

$$(x + z)(y + z) = xy + z$$

Figure 11.18(d) results.

Exercises

1. Starting with the circuit of Figure 11.12, produce an alternative circuit to that of Figure 11.13 which will carry out the same switching function.

2. A switching network has the connection matrix

$$\begin{bmatrix} 1 & x & 0 & (x+y) \\ x & 1 & y & xy \\ 0 & y & 1 & xy \\ (x+y) & xy & xy & 1 \end{bmatrix}$$

   Form the corresponding terminal matrix.

3. If nodes 3 and 4 of the network of Question 2 above are the only input–output nodes, derive the corresponding input–output matrix.

4. Using the determinantal approach, justify the star–delta and delta–star transformations given in the text.

5. If

$$X = \begin{bmatrix} 0 & 0 & 1 & 1 \\ 0 & 1 & 0 & 0 \\ 1 & 0 & 0 & 0 \\ 0 & 0 & 0 & 1 \end{bmatrix}, \quad Y = \begin{bmatrix} 1 & 0 & 1 & 0 & 0 & 1 & 0 & 1 \\ 1 & 1 & 0 & 0 & 0 & 0 & 0 & 1 \\ 0 & 0 & 1 & 1 & 0 & 1 & 0 & 1 \\ 0 & 0 & 0 & 0 & 0 & 0 & 0 & 0 \end{bmatrix}$$

   and

$$Z = \begin{bmatrix} 1 & 0 & 0 & 0 & 0 & 1 & 0 & 0 \\ 0 & 0 & 0 & 0 & 1 & 0 & 1 & 0 \\ 0 & 0 & 1 & 0 & 1 & 0 & 1 & 0 \\ 0 & 0 & 0 & 1 & 0 & 0 & 0 & 1 \end{bmatrix}$$

show that

   (i) $X(Y + Z) = XY + XZ$

   (ii) $X(YZ^T) = (XY)Z^T$

   (iii) $IZ = Z$

   (iv) $(XY)^T = Y^T X^T$

Chapter 12
Logic Equations

Introduction

One type of problem which occurs in logic design is that of producing the additional circuitry which will enable *existing* stock to be used, especially if large quantities of the existing circuits are available.

Example 1. Two existing circuits, which generate p and q from x, y, and z, are shown in Figure 12.1. Is it possible to construct three circuits, taking their

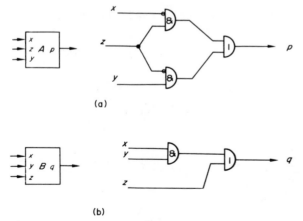

(a)

(b)

Figure 12.1

inputs from the set of logical variables a, b, and c and producing the x, y, and z for the above circuits, such that

$$p = \bar{a}b\bar{c} + \bar{b}c$$

and

$$q = ac + b?$$

This problem reduces to solving for x, y, and z, in terms of a, b, and c, the two simultaneous logic equations

$$\bar{x}z + y\bar{z} = \bar{a}b\bar{c} + \bar{b}c$$

and

$$xy + z = ac + b$$

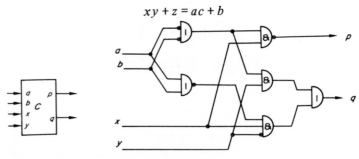

Figure 12.2

Example 2. Figure 12.2 is the diagram of a circuit which is already in existence. What values of x and y, as functions of a and b, will yield

$$p = a\bar{b}c + \bar{a}bc + \bar{a}\bar{b}\bar{c}$$

and

$$q = \bar{a}bc + \bar{a}b\bar{c} + abc?$$

This problem reduces to solving for x and y the two simultaneous logic equations

$$\overline{x(\bar{a} + \bar{b})} = a\bar{b}c + \bar{a}bc + \bar{a}\bar{b}\bar{c}$$

and

$$\bar{a}\bar{b}x\bar{y} + y(\bar{a} + \bar{b}) = \bar{a}bc + \bar{a}b\bar{c} + abc$$

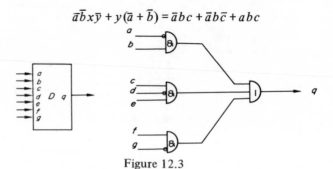

Figure 12.3

Example 3. A seven-input single-output circuit is shown in Figure 12.3, the input—output equation being

$$q = \bar{a}b + c\bar{d}e + f\bar{g}$$

Is it possible to utilize this circuit to generate $x'(= \bar{p}x + pyz)$, $y'(= \bar{p}y + p\bar{y}z + y\bar{z})$ and $z'(= \bar{p}z + p\bar{x}\bar{z})$?

The reformulation of this problem involves the solution of three distinct logic equations—one for each usage of the basic circuit—as follows:

(i) $\bar{a}b + c\bar{d}e + f\bar{g} = px + p\bar{x}yz$

(ii) $\bar{a}b + c\bar{d}e + f\bar{g} = \bar{p}y + p\bar{x}\bar{y}z + p\bar{x}yz$

(iii) $\bar{a}b + c\bar{d}e + f\bar{g} = \bar{p}z + p\bar{x}\bar{z}$

Incidently, this problem arises in Chapter 14 when Figure 14.17 is transformed into Figure 14.19.

The Standard Form of Logic Equations

Since any logic expression can be forced into the form of a sum of products, it follows that a logic equation can be made to assume the form of one sum of products equal to another. Therefore, there appears to be a fairly strong parallel between simultaneous linear algebraic equations and simultaneous logic equations. This similarity is purely superficial and a knowledge of the techniques for solving algebraic equations is useless when the solutions to a set of logic equations are required.

The essential difference between the number of solutions of a set of linear algebraic equations and a set of logic equations is that with the former it is possible to determine this number by inspection, whilst with the latter such an estimate is *not* possible. For example, the pair of simultaneous linear equations

$$ax + by = c$$

and

$$dx + ey = f$$

has *one* real value for x and *one* for y providing $ae \neq bd$.

This *a priori* knowledge is denied the logic designer. All that can be deduced, from an inspection of the set of logic equations to be solved, is the maximum number of solutions which may be involved. If there are n known variables, then the basis consists of 2^n columns and the maximum number of solutions, for each variable is 2^{2^n}. Obviously, when n is 2 the maximum number of solutions is 16, when n is 3 this becomes 256 $(= 16^2)$ and when n is 4 the maximum number of solutions reaches 65 536 $(= 256^2)$, i.e. increasing n by unity causes the maximum number of solutions to be squared. However, it is more important to realize that *there need not be any solutions at all* since a given equation can quite easily be insoluble.

For example, solve for x in terms of a the two distinct equations

(i) $a \rightarrow x = a$

(ii) $a \rightarrow x = \bar{a}$

The first equation has *no solutions* whereas the second has *two*, including the trivial case $x = 0$ (the other is $x = \bar{a}$).

Another important point of difference lies in the fact that, in the case of linear algebraic equations, m unknowns require m equations for a unique solution whereas with logic equations m unknowns may be determined for any number of equations from one upwards.

Therefore, any attempt at solving either a single logic equation, or a set of logic equations, must automatically involve an investigation of the *existence* of solutions. Moreover, if solutions exist, *all* the alternatives must be determined in order that the logic designer can make a choice of the most advantageous.

The procedure to be described will consider every possible combination of the known input variables, i.e. it will work from the basis formed from these variables. Either it will produce the designation number, with respect to this basis, of each unknown by generating the correct value corresponding to each column of the basis or else it will indicate that a designation number cannot be found. Since designation numbers are going to be involved, any equation such as

$$a + x = 1$$

or

$$a + x = 0$$

i.e. an equation which involves 1 or 0, will have the 1 replaced by an I (the designation number of I consists solely of 1s) and the 0 replaced by an O (the designation number of O consists solely of 0s). The above two equations would then become

$$a + x = I$$

and

$$a + x = O$$

It is possible to show that the first has solutions $x = a$ or $x = I$ whilst the second has no solutions at all.

Formulation of Equations

The first step is to re-write both sides, of each equation involved, as a sum of products. The elements of each product can be complicated logical functions of *either* the knowns *or* the unknowns but may *not* be a mixture of knowns and unknowns. For example, if a and b are the knowns and x and y the unknowns then the equation

$$a\bar{x} + \overline{b \rightarrow y} = a \equiv y$$

can be written in the form

$$\bar{x}a + \bar{y}b = ya + \bar{y}\bar{a}$$

Note that in any product the unknowns will be on the left and the knowns on the right. This is merely to maintain a consistent lay-out and is not essential. As another example, the equation

$$\bar{a} + (x \rightarrow y) + (a \equiv (b \equiv x)) = \bar{a}$$

can be written as

$$I\bar{a} + (x \rightarrow y)I + x(a \equiv b) + \bar{x}(a \not\equiv b) = I\bar{a}$$

Since the aim is to form a logical sum of products, this is achieved by taking every logical expression which contains either knowns only, or unknowns only, and forming the logical product of this expression with I. Although I is strictly *known* it can, in this method of solving equations, replace an unknown if necessary to ensure that a product exists. To distinguish between an I representing a known and one representing an unknown the symbols I_k and I_u will be employed. However, it must be realized that *both* these symbols represent quantities which have designation numbers consisting only of 1s.

With this convention, the last equation becomes

$$I_u\bar{a} + (x \rightarrow y)I_k + x(a \equiv b) + \bar{x}(a \not\equiv b) = I_u\bar{a}$$

Transformation to Matrix Form

The above equation can be written as

$$[I_u, (x \rightarrow y), x, \bar{x}] \times \begin{bmatrix} \bar{a} \\ I_k \\ a \equiv b \\ a \not\equiv b \end{bmatrix} = [I_u] \times [\bar{a}]$$

This will *always* be carried out, i.e. the original logic equation will be replaced by its matrix equivalent. Both sides will consist of a row matrix, having unknowns as elements, post-multiplied by a column matrix, having knowns as elements.

For example, given a known quantity a find x such that

$$a + x = 1$$

Using

$$a + x = I$$

it follows that

$$I_u a + x I_k = I_u I_k$$

from which

$$[I_u, x] \times \begin{bmatrix} a \\ I_k \end{bmatrix} = [I_u] \times [I_k]$$

can be formed. The problem has now been transformed to that of solving a logical matrix equation.

Insertion of Designation Numbers

Given the matrices, every element of which is a logical variable or combination of such variables, the next stage is to replace these matrices by matrices involving 0s and 1s only. This is done via the designation numbers. Two bases are used; one, the basis of the knowns, written as rows, produces the designation numbers which replace the elements of the column matrices and the other, the basis formed by the unknowns, written as columns, produces the designation numbers which replace the elements of the row matrices. An example will illustrate the point.

Consider the equation

$$\bar{x}a + y = \bar{b}$$

By the method described so far, this can be replaced by

$$[\bar{x}, y] \times \begin{bmatrix} a \\ I_k \end{bmatrix} = [I_u] \times [\bar{b}]$$

Table 12.1

a	0	1	0	1
b	0	0	1	1
I_k	1	1	1	1
$\bar{b}$	1	1	0	0

Table 12.2

x	y	I_u	$\bar{x}$
0	0	1	1
0	1	1	1
1	0	1	0
1	1	1	0

Using the basis $b[a, b]$ in row form, Table 12.1 results and using the basis $b[y, x]$ in column form gives Table 12.2. Hence the equation becomes

$$\begin{bmatrix} 1 & 0 \\ 1 & 1 \\ 0 & 0 \\ 0 & 1 \end{bmatrix} \times \begin{bmatrix} 0 & 1 & 0 & 1 \\ 1 & 1 & 1 & 1 \end{bmatrix} = \begin{bmatrix} 1 \\ 1 \\ 1 \\ 1 \end{bmatrix} \times \begin{bmatrix} 1 & 1 & 0 & 0 \end{bmatrix}$$

which reduces to

$$\begin{bmatrix} 0 & 1 & 0 & 1 \\ 1 & 1 & 1 & 1 \\ 0 & 0 & 0 & 0 \\ 1 & 1 & 1 & 1 \end{bmatrix} = \begin{bmatrix} 1 & 1 & 0 & 0 \\ 1 & 1 & 0 & 0 \\ 1 & 1 & 0 & 0 \\ 1 & 1 & 0 & 0 \end{bmatrix}$$

Existence and Determination of Solutions

The decision now to be faced is that of interpreting the equality between the two matrices representing the left- and right-hand sides of the equation. Since these matrices represent every possible way in which 0 and 1 can be ascribed to both the knowns and the unknowns, an equation in m unknowns and n knowns will result in a $2^m \times 2^n$ matrix. Hence, if the values in corresponding positions of both matrices *are the same*, then the values of the knowns and unknowns which produced those identical values will be *valid*, i.e. if the operation of equivalence is carried out element by element on these matrices, the resulting matrix will indicate the valid combinations. Therefore replacing the equals by EQUIVALENCE and evaluating, the above example produces the resulting matrix

$$\mathbf{A} = \begin{bmatrix} 0 & 1 & 1 & 0 \\ 1 & 1 & 0 & 0 \\ 0 & 0 & 1 & 1 \\ 1 & 1 & 0 & 0 \end{bmatrix}$$

where the element a_{rs} refers to the combination due to row r of the basis of the unknowns and column s of the basis of the knowns.

In the above example, it can be seen that, for row 0 with respect to the basis $b[y, x]$, the only valid positions are in columns 1 and 2 of the basis $b[a, b]$, i.e. $x = 0$ and $y = 0$ in the designation numbers of the solutions for x and y in positions 1 and 2 with respect to the basis $b[a, b]$. This is indicated in Table 12.3. Further for row 1 with respect to the basis $b[y, x]$, the only valid positions correspond to columns 0 and 1 and Table 12.4 results.

However, an interesting point now arises, namely the discrepancy between the values for x and y in column 1. This is explained by the fact that more than

Table 12.3

a	0	1	0	1
b	0	0	1	1
x	–	0	0	–
y	–	0	0	–

Table 12.4

a	0	1	0	1
b	0	0	1	1
x	0	0	–	–
y	1	1	–	–

one solution exists and Tables 12.3 and 12.4 can be replaced by Tabio 12.5. By similar reasoning for row 2, Table 12.6 and hence Table 12.7 result. Finally for row 3, Table 12.8 and hence Table 12.9 yield the totality of solutions, namely 12 sets in all.

Table 12.5

	a	0	1	0	1
	b	0	0	1	1
1st set	x	0	0	0	–
	y	1	0	0	–
2nd set	x	0	0	0	–
	y	1	1	0	–

Table 12.6

a	0	1	0	1
b	0	0	1	1
x	–	–	1	1
y	–	–	0	0

Table 12.7

	a	0	1	0	1
	b	0	0	1	1
1st set	x	0	0	0	1
	y	1	0	0	0
2nd set	x	0	0	0	1
	y	1	1	0	0
3rd set	x	0	0	1	1
	y	1	0	0	0
4th set	x	0	0	1	1
	y	1	1	0	0

Table 12.8

a	0	1	0	1
b	0	0	1	1
x	1	1	–	–
y	1	1	–	–

The total number of possible solution combinations can quickly be deduced from a consideration of the number of 1s in the columns of the result matrix. In this example there are 2 in column 0, 3 in column 1, 2 in column 2, and 1 in column 3. Therefore, the total number of solutions is 2 x 3 x 2 x 1 (= 12). Note that, if *any* column has no 1s, then *no solutions* can be found.

To determine a valid solution of the equation, any one of the pairs of designation numbers can be translated into a logical function of the unknowns.

Table 12.9

a	0	1	0	1
b	0	0	1	1
x	0	0	0	1
y	1	0	0	0
x	0	0	0	1
y	1	1	0	0
x	0	0	1	1
y	1	0	0	0
x	0	0	1	1
y	1	1	0	0
x	1	0	0	1
y	1	0	0	0
x	1	0	0	1
y	1	1	0	0
x	1	0	1	1
y	1	0	0	0
x	1	0	1	1
y	1	1	0	0
x	0	1	0	1
y	1	1	0	0
x	0	1	1	1
y	1	1	0	0
x	1	1	0	1
y	1	1	0	0
x	1	1	1	1
y	1	1	0	0

For example, the fourth set give the result that

$$x = b \quad \text{and} \quad y = \bar{b}$$

is a pair of solutions to the original equation

$$\bar{x}a + y = \bar{b}$$

as can easily be verified.

Similarly, the first set shows that

$$x = ab$$

and

$$y = \bar{a}\bar{b}$$

is another pair of solutions. This can be checked by substitution as follows:

$$
\begin{aligned}
\text{L.H.S.} &= (\overline{ab})a + \bar{a}\bar{b} \\
&= (\bar{a} + \bar{b})a + \bar{a}\bar{b} \\
&= a\bar{b} + \bar{a}\bar{b} \\
&= (a + \bar{a})\bar{b} \\
&= \bar{b} \\
&= \text{R.H.S.}
\end{aligned}
$$

Note that all checks should be carried out on the *original* equations.

Simultaneous Logical Equations

For more than one logic equation, the approach is initially identical, i.e. consider each equation separately and produce its result matrix as if it were the *only* equation. The sole restriction is that the same basis *must* be used for all the equations, i.e. if any of the variables (known or unknown) are missing from any one of the equations this does not mean that the basis used for that equation can be truncated.

For example, suppose that in addition to the equation

$$\bar{x}a + y = \bar{b}$$

the equation

$$x\bar{a} = y$$

must also hold.

In the second equation the *only* known is a but the full basis $b[a, b]$ must still be used as follows:

$$xa = y$$
$$xa = yI_k$$
$$[x] \times [\bar{a}] = [y] \times [I_k]$$

$$\begin{bmatrix} 0 \\ 0 \\ 1 \\ 1 \end{bmatrix} \times [1 \ 0 \ 1 \ 0] = \begin{bmatrix} 0 \\ 1 \\ 0 \\ 1 \end{bmatrix} \times [1 \ 1 \ 1 \ 1]$$

$$\begin{bmatrix} 0 & 0 & 0 & 0 \\ 0 & 0 & 0 & 0 \\ 1 & 0 & 1 & 0 \\ 1 & 0 & 1 & 0 \end{bmatrix} = \begin{bmatrix} 0 & 0 & 0 & 0 \\ 1 & 1 & 1 & 1 \\ 0 & 0 & 0 & 0 \\ 1 & 1 & 1 & 1 \end{bmatrix}$$

Replacing the equals by EQUIVALENCE, the result matrix for the second equation is

$$\begin{bmatrix} 1 & 1 & 1 & 1 \\ 0 & 0 & 0 & 0 \\ 0 & 1 & 0 & 1 \\ 1 & 0 & 1 & 0 \end{bmatrix}$$

i.e. there are 2 x 2 x 2 x 2 (= 16) solutions. However, the twelve solutions for the 1st equation and the sixteen solutions of the second only have one in common.

The procedure to determine the *common solutions* is merely to form the logical product of each of the result matrices, element by element. In this case

$$\begin{bmatrix} 0 & 1 & 1 & 0 \\ 1 & 1 & 0 & 0 \\ 0 & 0 & 1 & 1 \\ 1 & 1 & 0 & 0 \end{bmatrix} \begin{bmatrix} 1 & 1 & 1 & 1 \\ 0 & 0 & 0 & 0 \\ 0 & 1 & 0 & 1 \\ 1 & 0 & 1 & 0 \end{bmatrix} = \begin{bmatrix} 0 & 1 & 1 & 0 \\ 0 & 0 & 0 & 0 \\ 0 & 0 & 0 & 1 \\ 1 & 0 & 0 & 0 \end{bmatrix}$$

Hence the pair of logic equations have only one solution in common. This is given by Table 12.10 and so the pair of solutions

$$x = a \equiv b$$

and

$$y = \overline{(a + b)}$$

is one form of the answer. The other possible forms of the answer are merely different interpretations of the same designation numbers.

Table 12.10

a	0	1	0	1
b	0	0	1	1
x	1	0	0	1
y	1	0	0	0

As a second example, solve for x and y in terms of a, b, and c the pair of simultaneous equations

$$\bar{y}a + y\bar{a}b = \bar{y}(b \to a) + ya$$

and

$$xb + yc = (x \equiv y)(a \equiv b) + (x + y)(a \not\equiv b)$$

Although the first equation does not contain the unknown x, these equations are *both* solved using the bases $b[a, b, c]$ and $b[y, x]$. The solution matrix for the first equation is

$$\begin{bmatrix} 0 & 1 & 1 & 1 & 0 & 1 & 1 & 1 \\ 1 & 0 & 0 & 0 & 1 & 0 & 0 & 0 \\ 0 & 1 & 1 & 1 & 0 & 1 & 1 & 1 \\ 1 & 0 & 0 & 0 & 1 & 0 & 0 & 0 \end{bmatrix}$$

and that of the second is

$$\begin{bmatrix} 0 & 1 & 1 & 0 & 0 & 1 & 1 & 0 \\ 1 & 0 & 0 & 1 & 0 & 1 & 1 & 0 \\ 1 & 0 & 1 & 0 & 1 & 0 & 1 & 0 \\ 0 & 0 & 1 & 1 & 1 & 1 & 1 & 1 \end{bmatrix}$$

Forming the logical product of the two result matrices yields

$$\begin{bmatrix} 0 & 1 & 1 & 0 & 0 & 1 & 1 & 0 \\ 1 & 0 & 0 & 0 & 0 & 0 & 0 & 0 \\ 0 & 0 & 1 & 0 & 0 & 0 & 1 & 0 \\ 0 & 0 & 0 & 0 & 1 & 0 & 0 & 0 \end{bmatrix}$$

The two columns of 0s indicate that there are no solutions for the equations *when they are treated as simultaneous equations* even though the first has 256 pairs of solutions and the second has 288 pairs of solutions when they are considered separately.

Constraints

It is possible to restrict the number of columns of any basis in order that a meaningful result be obtained. For example, in the above matrix the two columns of 0s indicated that no solutions existed *unless* the combinations of the elements of the basis, which generated those particular columns, could *never* occur, i.e. if the pair of values $a = 1$ and $b = 1$ could never co-exist. Under these circumstances, the basis b [a, b, c] reduces to the *constrained basis* defined by

$$\# a = 0 \quad 1 \quad 0 \quad 0 \quad 1 \quad 0$$
$$\# b = 0 \quad 0 \quad 1 \quad 0 \quad 0 \quad 1$$
$$\# c = 0 \quad 0 \quad 0 \quad 1 \quad 1 \quad 1$$

There is little point in writing down the columns of combinations which will never occur.

Table 12.11

z	0	1	0	1	0	1	0	1
y	0	0	1	1	0	0	1	1
x	0	0	0	0	1	1	1	1
f	−	0	1	0	1	−	1	0
g	−	1	1	1	0	−	1	1

Some method of indicating the constraints, which must be placed upon the elements of the basis, is required. For example, consider Table 12.11. The meaningful columns are 1, 2, 3, 4, 6, and 7, those which are to be ignored being indicated with dashes. It is necessary to be able to determine, by means of some functional relationship, which columns are to be used and which are to be ignored. This functional relationship will be defined by a logic equation, or set of equations, involving the elements of the basis and *I* (universally true) or *O* (universally false). The general form of these equations will be either a sum of products which is equal to *I*, a product of sums which is equal to *I*, a sum of products which is equal to *O*, or a product of sums which is equal to *O*. Each of these must be capable of representing the same constraint and it will be a matter of convenience which is used.

The constraint is first written as a designation number with respect to the basis in question. In this designation number, a 1 will be used to indicate that the column above it is to be retained as being meaningful whilst a 0 will be used to indicate that the column above it in the basis refers to a combination of elements of the basis which can never occur. Table 12.12 shows how the constraint for Table 12.11 is indicated.

Table 12.12

z	0	1	0	1	0	1	0	1
y	0	0	1	1	0	0	1	1
x	0	0	0	0	1	1	1	1
C	0	1	1	1	1	0	1	1

The logic equation defining the constraint can then be formulated in one of the four ways mentioned above. In the case of Table 12.12, the constraint C is given by

$$C : x\bar{y}\bar{z} + \bar{x}\bar{y}z + xy\bar{z} + \bar{x}y\bar{z} + \bar{x}yz + xyz = I$$

or

$$C : (x + y + z)(\bar{x} + y + \bar{z}) = I$$

or

$$C : \bar{x}\bar{y}\bar{z} + x\bar{y}z = O$$

or

$$C : (\bar{x} + y + z)(x + y + z)(\bar{x} + \bar{y} + z)(x + \bar{y} + z)(x + \bar{y} + \bar{z})(\bar{x} + \bar{y} + \bar{z}) = O$$

In all four equations, the values of x, y, and z which satisfy the equations select only the six columns of the constrained basis required. The above forms of the constraints can themselves be simplified to

$$C : y + (\bar{x}z + x\bar{z}) = I$$
$$C : y + (x + z)(\bar{x} + \bar{z}) = I$$
$$C : \bar{y}(x + \bar{z})(\bar{x} + z) = O$$

and

$$C : \bar{y}(\bar{x}\bar{z} + xz) = O$$

Note the dual nature of the above and preceding constraint equations. Given one constraint equation in terms of I (say), then another equally valid equation results from negating all the terms, including the I, and interchanging the logical connectives AND and OR whenever they occur.

The expressions for f and g in terms of the unconstrained basis of Table 12.11 are

$$f = (y + x)\bar{z}$$

and

$$g = y + \bar{x}z$$

However, when the constrained basis of Table 12.13 is used, the expressions for f and g simplify to

$$f = \bar{z}$$

and

$$g = y + z$$

Table 12.13

z	1	0	1	0	0	1
y	0	1	1	0	1	1
x	0	0	0	1	1	1
f	0	1	0	1	1	0
g	1	1	1	0	1	1

The circuit requirements being reduced from two OR and two AND gates to a single OR gate. Any improvement which takes place arises from the fact that either logical value can be inserted in the blank columns. In terms of the Karnaugh map, the corresponding locations can be treated as 'don't care'

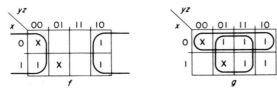

Figure 12.4

conditions. The simplification by means of Karnaugh maps is shown in Figure 12.4, where X represents either 1 or 0 which ever is the most convenient. In the case of g shown, X is assumed to be 1 in each case.

Multiple Constraints

If there exist a number of constraints, then the overall constraint is obtained by forming their logical product. It is often easier to see the pattern if all the constraints are expressed in terms of I or, alternatively, all in terms of O.

For example, if the standard basis b$[z, y, x]$ is subject to the three constraints

$$C_1 : \ x + y = I$$
$$C_2 : \ \bar{y} + z = I$$

and

$$C_3 : \ xyz = O$$

find the overall constraint C. Since $xyz = 0$ is the same as $\overline{xyz} = I$, i.e. $\bar{x} + \bar{y} + \bar{z} = I$,

$$C_3 : \ \bar{x} + \bar{y} + \bar{z} = I$$

Thus

$$C : (x + y)(\bar{y} + z)(\bar{x} + \bar{y} + \bar{z}) = I$$

Alternatively,

$$C_1 : \ \bar{x}\bar{y} = O$$

and

$$C_2 : \ y\bar{z} = O$$

therefore

$$C : \bar{x}\bar{y} + y\bar{z} + xyz = O$$

With respect to the standard basis, both these representations of C yield the designation number 0 0 0 1 1 1 0 0. The basis and the constraints are shown in Table 12.14 and the contrained basis in Table 12.15.

With such a constrained basis, considerable logic simplification can be expected. One immediate consequence is that, for this particular overall constraint, $x = \bar{y}$, i.e. the three variables x, y, and z are no longer *independent*—in fact only two are necessary, namely x and z or y and z.

Table 12.14

z	0	1	0	1	0	1	0	1
y	0	0	1	1	0	0	1	1
x	0	0	0	0	1	1	1	1
C_1	0	0	1	1	1	1	1	1
C_2	1	1	0	1	1	1	0	1
C_3	1	1	1	1	1	1	1	0
C	0	0	0	1	1	1	0	0

Table 12.15

z	1	0	1
y	1	0	0
x	0	1	1
$\bar{x}$	1	0	0

Example. Simplify

$$f = xy\bar{z} + \bar{x}\bar{y}\bar{z} + yz$$

and

$$g = xyz + x\bar{y}z + \bar{x}yz$$

subject to the three constraints C_1, C_2, and C_3 above. The constrained basis is given by Table 12.15 as before and the designation numbers of f and g, with respect to the constrained basis, are 1 0 0 and 1 0 1 respectively. Hence, $f = y$ (or $\bar{x}$) and $g = z$.

Logical Dependence

The question of the independence of logic functions can now be considered in general. Independence is defined as follows. Two, or more, logic functions are *logically independent* if every possible combination of truth values occurs when their designation numbers are written in full.

For example, $x \equiv y$ and $\bar{x}$ are logically independent because

$$\#y = 0 \quad 1 \quad 0 \quad 1$$
$$\#x = 0 \quad 0 \quad 1 \quad 1$$

$$\#x \equiv y = 1 \quad 0 \quad 0 \quad 1$$
$$\#\bar{x} = 1 \quad 1 \quad 0 \quad 0$$

and it can be seen that all four columns of these two functions are distinct, i.e. every possible pair of logical values occur.

On the other hand, $y \to x$ and y are not independent since

$$
\begin{array}{lcccc}
\#y = 0 & 1 & 0 & 1 \\
\#x = 0 & 0 & 1 & 1 \\
\hline
\#y \to x = 1 & 0 & 1 & 1 \\
\#\bar{y} = 1 & 0 & 1 & 0
\end{array}
$$

from which it can be seen that only three of the four possible combinations are present. With n variables there can be at most only n independent equations.

Consider $x \equiv y$, $y \to x$, and $\bar{y}$. How can the dependence or otherwise be determined? That there must be a dependent relationship in this case is obvious since there are three functions of two variables. For convenience the three functions will be referred to as f, g, and h, where

$$f = x \equiv y$$

$$g = y \to x$$

and

$$h = \bar{y}$$

and, using the standard basis $b[y, x]$, Table 12.16 results. Re-ordering the columns of f, g, and h into standard basis order, leaving blanks where necessary, produces Table 12.17. This is obviously constrained and the

Table 12.16

y	0	1	0	1
x	0	0	1	1
f	1	0	0	1
g	1	0	1	1
h	1	0	1	0

Table 12.17

f	0	–	–	1	–	–	0	1
g	0	–	–	1	–	–	1	1
h	0	–	–	0	–	–	1	1

constraint C is given by the designation number 1 0 0 1 0 0 1 1 *with respect to the re-arranged basis.* Note that three variables f, g, and h imply 2^3 columns. In this example

$$C : g(f + h) + \bar{f}\bar{g}\bar{h} = I$$

and this is the dependent relationship.

Table 12.17 has a deficiency of possible values but there are no duplicate entries as was the case with $y \to x$ and $\bar{y}$ only. Let $f = y \to x$ and $g = \bar{y}$. Then, as before

$$
\begin{array}{lcccc}
\#y = 0 & 1 & 0 & 1 \\
\#x = 0 & 0 & 1 & 1 \\
\hline
\#f = 1 & 0 & 1 & 1 \\
\#g = 1 & 0 & 1 & 0
\end{array}
$$

where the deficiency is due to the omission of $f = 0$ and $g = 1$ and the replacement of this pair of logical values by $f = g = 1$. In such cases, re-order the columns of f and g *ignoring any duplicate columns*, i.e.

$$
\begin{array}{cccc}
f & 0 & 1 & -1 \\
g & 0 & 0 & -1
\end{array}
$$

from which it can be seen that the constraint has the designation number 1 1 0 1 with respect to the basis b$[f, g]$ giving

$$C : \bar{g} + f = I$$

The validity of this should be checked by substitution. For example,

$$
\begin{aligned}
\bar{g} + f &= \bar{\bar{y}} + (y \to x) \\
&= y + \bar{y} + x \\
&= I + x \\
&= I
\end{aligned}
$$

Example. Three logic circuits have been constructed. The first has three inputs a, b, and c and one output g such that

$$g = (a \equiv b) \to c$$

The second has two inputs b and c and one output h such that

$$h = b \to \bar{c}$$

and the third has three inputs g, h, and x and one output z such that

$$z = xgh$$

Show that it is only possible to produce a logic circuit, with a, b, and c as inputs and z as outputs, such that

$$z = a(b \not\equiv c) + \bar{a}b$$

if the basis b$[a, b, c]$ is constrained in some fashion.

This is equivalent to solving, for x in terms of a, b, and c, the equation

$$x(((a \equiv b) \to c)(b \to \bar{c})) = a(b \not\equiv c) + \bar{a}b$$

and determining any constraints necessary for the solution.

$$[x] \times [((a \equiv b) \to c)(b \to c)] = [I_u] \times [a(b \not\equiv c) + \bar{a}b]$$

and, using the basis of Table 12.18, the equation can be written, in matrix form, as

$$
\begin{bmatrix} 0 \\ 1 \end{bmatrix} \times [0 \ \ 0 \ \ 1 \ \ 1 \ \ 0 \ \ 1 \ \ 1 \ \ 0] = \begin{bmatrix} 1 \\ 1 \end{bmatrix} \times [0 \ \ 1 \ \ 0 \ \ 0 \ \ 1 \ \ 1 \ \ 1 \ \ 1]
$$

$$\begin{bmatrix} 0 & 0 & 0 & 0 & 0 & 0 & 0 & 0 \\ 0 & 0 & 1 & 1 & 0 & 1 & 1 & 0 \end{bmatrix} = \begin{bmatrix} 0 & 1 & 0 & 0 & 1 & 1 & 1 & 1 \\ 0 & 1 & 0 & 0 & 1 & 1 & 1 & 1 \end{bmatrix}$$

which yields the result matrix

$$\begin{bmatrix} 1 & 0 & 1 & 1 & 0 & 0 & 0 & 0 \\ 1 & 0 & 0 & 0 & 0 & 1 & 1 & 0 \end{bmatrix}$$

Obviously, there are no solutions possible unless the three columns consisting only of 0s can be ignored. Hence, with respect to the basis $b[a, b, c]$ the

Table 12.18

a	0	1	0	1	0	1	0	1
b	0	0	1	1	0	0	1	1
c	0	0	0	0	1	1	1	1
$a(b \not\equiv c) + \bar{a}b$	0	0	1	1	0	1	1	0
g	0	1	1	0	1	1	1	1
h	1	1	0	0	1	1	1	1
gh	0	1	0	0	1	1	1	1

x	I_u
0	1
1	1

(a)   (b)

constraint must have a designation number 1 0 1 1 0 1 1 0, i.e. the constraint is given by

$$C : (\bar{a} + b + c)(a + b + \bar{c})(\bar{a} + \bar{b} + \bar{c}) = I$$

In this case, there are two solutions shown, with reference to the constrained basis, in Table 12.19. Thus the logic expression for x is either c or $c + \bar{b}$, with

Table 12.19

a	0	0	1	1	0
b	0	1	1	0	1
c	0	0	0	1	1
1st solution	1	0	0	1	1
2nd solution	0	0	0	1	1

respect to the constrained basis. Note that, once the constraint is applied, there will be sixteen solutions for x since each of the missing three columns will contribute two alternatives.

In the introduction to this chapter, the first example used to justify the need for the systematic approach which followed required the determination of x, y, and z in terms of a, b, and c where

$$\bar{x}z + y\bar{z} = \bar{a}b\bar{c} + \bar{b}c$$

and

$$xy + z = ac + b$$

Using the matrix approach developed above, it can be shown that there are 432 distinct sets of solutions to the problem. Choosing the set given by

$$\#a = 0 \quad 1 \quad 0 \quad 1 \quad 0 \quad 1 \quad 0 \quad 1$$
$$\#b = 0 \quad 0 \quad 1 \quad 1 \quad 0 \quad 0 \quad 1 \quad 1$$
$$\#c = 0 \quad 0 \quad 0 \quad 0 \quad 1 \quad 1 \quad 1 \quad 1$$

$$\#x = 0 \quad 0 \quad 0 \quad 1 \quad 0 \quad 1 \quad 0 \quad 0$$
$$\#y = 0 \quad 0 \quad 1 \quad 1 \quad 1 \quad 1 \quad 0 \quad 0$$
$$\#z = 0 \quad 0 \quad 1 \quad 1 \quad 0 \quad 0 \quad 1 \quad 1$$

then the functional forms of x, y, and z are given by

$$x = ay$$
$$y = b\bar{c} + \bar{b}c$$

and

$$z = b$$

These require three AND gates and one OR gate to generate x, y, and z from a, b, and c and the saving hardly seems worth while. However, a careful look at the circuits of Figure 12.1 show that these same circuits can be used to form the required values providing the 'z' input of the B circuit is set to 0, i.e. it is not connected. Figure 12.5 shows the solution.

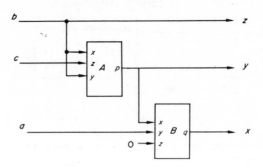

Figure 12.5

The second example poses the problem of solving, for x and y in terms of a and b, the two simultaneous equations

$$x(\overline{\overline{a} + \overline{b}}) = a\overline{b}c + \overline{a}bc + \overline{a}\overline{b}\overline{c}$$

and

$$\overline{a}\overline{b}x\overline{y} + y(\overline{a} + \overline{b}) = \overline{a}bc + \overline{a}b\overline{c} + abc$$

The matrix techniques developed in this chapter *prove* that no solutions exist and so the logic designer cannot utilize the already existing circuit of Figure 12.2.

Example 3 involves solving three independent logic equations, each involving four knowns and seven unknowns. One of the many possible sets of solutions is shown in Figure 14.19 of Chapter 14.

Suggestions for Further Reading

Ledley, R. S., 1960, *Digital Computer and Control Engineering*, McGraw-Hill, New York.

Exercises

1. Justify the statements, made in the text, that
    (i) $a \to x = a$ has *no* solutions,
    (ii) $a \to x = \overline{a}$ has two solutions, namely $x = 0$ and $x = \overline{a}$.

2. Show that the equation

$$ax + (\overline{x} \to b) = I$$

has a solution for x which is independent of a and find this solution explicitly.

3. Solve for r in terms of p and q the logic equation

$$(r \equiv \overline{p})(\overline{r} \not\equiv q) = 0$$

4. Solve for a and b in terms of r and s

$$as + b = r \equiv s$$

5. Solve for x and y in terms of a, b, and c the following simultaneous logic equations:

$$xb + y(a + c) = (\bar{a} \rightarrow c)b$$
$$x = y$$

6. Given the simultaneous logic equations

$$(x \rightarrow b) + yc = \bar{x}$$
$$xy + \bar{x} = a \equiv b$$

Determine how many solutions exist
   (i) for x and y in terms of a, b, and c,
   (ii) for a, b, and c in terms of x and y.

7. Show that the logic equation

$$x\bar{a}\bar{b} + \bar{x}(a + b) = \bar{x}(a \equiv b)$$

has *no* solutions for x in terms of a and b unless the latter are constrained to satisfy $(a + b) = I$.
   Find all the solutions when this constraint is satisfied.

8. Show that

$$y + \bar{c} = (\overline{ac}) + b$$

has 8^4 solutions for y, in terms of a, b, and c, whilst

$$xy = abc$$

has 3^7 solutions for x and y, in terms of a, b, and c.
   Show further that when they are considered as simultaneous equations there are 162 solutions only.

9. With the data of Question 8 above, namely

$$y + \bar{c} = (\overline{ac}) + b$$

and

$$xy = abc$$

   (i) if $x = a$ is one solution for x, find *all* the corresponding solutions for y
   (ii) if, in addition to the two equations given, it is required to have $x = y$ show that the constraint

$$C : a + \bar{c} = I$$

is necessary.

Chapter 13
Basic Arithmetical Units

Introduction

Since the operations of multiplication and division can be reduced to addition and subtraction, only adders and subtractors need be considered as being essential. If complementation circuits are allowed, then an adder is the only other basic circuit necessary. A study of binary adders and subtractors is, therefore, essential to the understanding of digital computer logic design. In order that special purpose equipment can be constructed, the problems involved in designing adders using coded decimal representations must also be considered. The use of these units in turn demands a knowledge of devices capable of carrying out the conversion from one representation to another and for the detection of invalid combinations of digits.

Adders

The fundamental elementary logic configuration is that of the *half-adder*. This device accepts as inputs two binary digits, which will be either 0 or 1, and generates a *sum* bit, if only one of them is a 1, and a *carry* bit, if they are both 1s, otherwise both the sum and carry bits are 0s. The truth table for such a device is given in Table 13.1 and its corresponding block schematic in Figure 13.1.

The Boolean expressions for s (sum) and c (carry) are not unique unless their special normal forms are written out in full. In general they will be most

Table 13.1

	$x + y$			
y	0	1	0	1
x	0	0	1	1
Sum	0	1	1	0
Carry	0	0	0	1

Figure 13.1

economically expressed symbolically as mongrel forms, i.e. any convenient mixture of operands and operators. Some of these expressions are listed in Tables 13.2 and 13.3 and their implementations are shown in Figures 13.2 and 13.3—the letters (a), (b), (c), etc., associated with the lines of the tables

Table 13.2

	Sum $(x + y)$
(a)	$x \not\equiv y$
(b)	$x\bar{y} + \bar{x}y$
(c)	$(x + y)(\bar{x} + \bar{y})$
(d)	$((xy) \downarrow (\bar{x}\bar{y}))$
(e)	$((x + \bar{y})\mid (\bar{x} + y))$

Table 13.3

	Carry $(x + y)$
(a)	xy
(b)	$\bar{x} \downarrow \bar{y}$
(c)	$\overline{(x \rightarrow \bar{y})}$

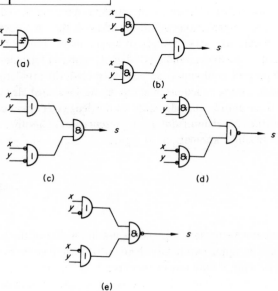

Figure 13.2

correspond to the letters of the figures. Note that in Figure 13.3(c) the expression $x \rightarrow y$ involves the convention of Chapter 5 and results is the same diagrammatic representation as Figure 13.3(b). An appeal can be made to

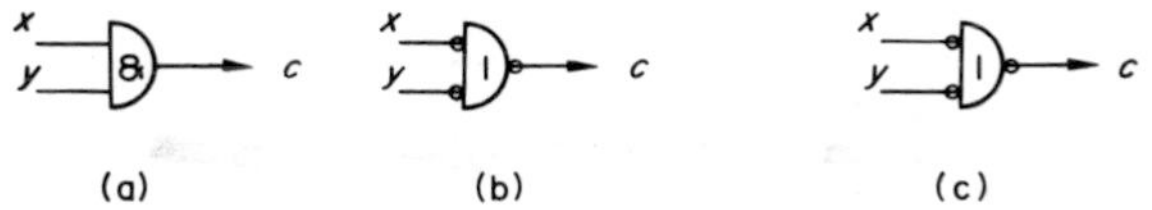

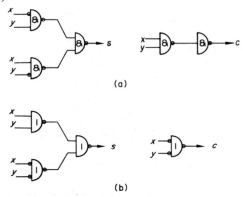

(a)　　　　　　(b)　　　　　　(c)

Figure 13.3

De Morgan's rules to enable s and c to be generated in terms of NOR or NAND elements only. For example, from Table 13.2

$$s = x\bar{y} + \bar{x}y$$

and hence, by De Morgan's rules,

$$s = (\bar{x}|y)\big|(x|\bar{y})$$

Similarly,

$$c = \overline{(x|y)}$$

Thus s and c have been expressed in terms of NAND elements only. These equations are realized by the circuits of Figure 13.4(a), and their NOR equivalents, namely

$$s = (x \downarrow y) \downarrow (\bar{x} \downarrow \bar{y})$$

and

$$c = \bar{x} \downarrow \bar{y}$$

by Figure 13.4(b).

(a)

(b)

Figure 13.4

In practice, when electronic elements are used, these implementations are extremely popular. This is particularly true of modern integrated silicon chip circuitry when a number of two or more input NOR (or NAND) gates are built

into a single chip. The production of a half-adder from a quadruple two-input
NOR chip is shown in Figure 13.5 where, as is often the case, the negated
values of the inputs are available. If these negated values are not available then

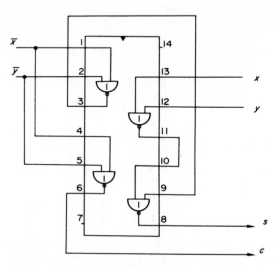

Figure 13.5

either a more complex chip, as in Figure 13.6, or more than one chip can be
used.

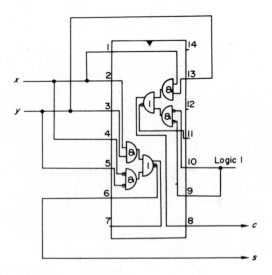

Figure 13.6

For fluidic elements, on the other hand, the NOT-IMPLIES element is easily constructed and for these devices the formulae for s and c are derived from

$$s = (x \rightarrow y) \rightarrow (\overline{\overline{x} \rightarrow \overline{y}})$$

and

$$c = \overline{(x \rightarrow \overline{y})}$$

Negation and implication can be obtained from the identities

$$\bar{a} = \overline{(1 \rightarrow a)}$$

and

$$a \rightarrow b = \overline{(1 \rightarrow (\overline{a \rightarrow b}))}$$

Hence, s and c can be expressed in terms of the inputs x and y, a logic level corresponding to 1 and fluidic NOT-IMPLIES devices. Whilst fluidic devices have been mentioned briefly here, no further attempt will be made to study their behaviour.

Full-adders

In a practical binary adder, the possibility that a carry bit from a previous stage must be taken into account gives rise to the necessity for designing a *full-adder*.

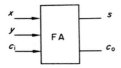

Figure 13.7

This a three-input, two-output device in which the inputs are x, y, and c_i (the *carry-in* from a previous stage) and the outputs are s and c_o (the *carry-out* to the next stage). Its block schematic is shown in Figure 13.7 and the corresponding truth table in Table 13.4. The description of its operation is as follows. A

Table 13.4

y	0	1	0	1	0	1	0	1
x	0	0	1	1	0	0	1	1
c_i	0	0	0	0	1	1	1	1
s	0	1	1	0	1	0	0	1
c_o	0	0	0	1	0	1	1	1

sum bit occurs if there is an odd number of input bits whilst a carry-out occurs if there are two or more input bits. Otherwise both the sum and the carry-out will be 0.

From the truth table it can be seen that

$$s = x\bar{y}\bar{c}_i + \bar{x}y\bar{c}_i + \bar{x}\bar{y}c_i + xyc_i$$

and

$$c_o = xy\bar{c}_i + x\bar{y}c_i + \bar{x}yc_i + xyc_i$$

Obviously, the expressions on the right-hand sides of these equations *must* be symmetrical in x, y, and c_i. Whilst the above special normal forms can easily be simplified by the usual techniques, it is easy to see, in these cases, that

$$s = (x \not\equiv y)\bar{c}_i + (x \equiv y)c_i$$

which reduces as follows:

$$s = (x \not\equiv y)\bar{c}_i + \overline{(x \not\equiv y)}c_i$$
$$= ((x \not\equiv y) \not\equiv c_i)$$
$$= x \not\equiv y \not\equiv c_i$$

Similarly

$$c_o = (xy)\bar{c}_i + (x + y)c_i$$
$$= xy\bar{c}_i + xc_i + yc_i$$
$$= xy\bar{c}_i + xyc_i + xc_i + yc_i$$

(since $a = a + b$ for all a and b). Hence

$$c_o = xy(\bar{c}_i + c_i) + (x + y)c_i$$
$$= xy + (x + y)c_i$$

Alternatively,

$$c_o = xy\bar{c}_i + xc_i + yc_i$$
$$= xy\bar{c}_i + xyc_i + x\bar{y}c_i + xyc_i + \bar{x}yc_i$$
$$= xy(\bar{c}_i + c_i) + x\bar{y}c_i + \bar{x}yc_i$$
$$= xy + (x \not\equiv y)c_i$$

The variations due to the permutation of the inputs are listed in Table 13.5, the last three being of particular importance as they indicate the method of

Table 13.5

s	$x \not\equiv y \not\equiv c_i$
c_o	$xy + (x + y)c_i$
	$yc_i + (y + c_i)x$
	$c_ix + (c_i + x)y$
	$xy + (x \not\equiv y)c_i$
	$yc_i + (y \not\equiv c_i)x$
	$c_ix + (c_i \not\equiv x)y$

using two half-adders to construct a full-adder. Let the first half-adder of Figure 13.8 have outputs $s_1 = x \not\equiv y$ and $c_1 = xy$ from inputs x and y, then the outputs of the second half-adder are

$$s_2 = c_i \not\equiv s_1$$

and

$$c_2 = c_i s_1$$

But

$$s = x \not\equiv y \not\equiv c_i$$

yielding

$$s = s_2$$

and

$$c_o = xy + (x \not\equiv y)c_i$$

yielding

$$c_o = c_1 + c_2$$

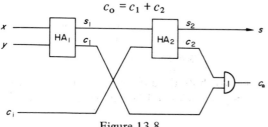

Figure 13.8

A full-adder constructed from NOR units only is shown in Figure 13.9. This particular representation generates $\bar{s}$ and $\bar{c}_o$ as well as s and c_o. In general, if a negated output can be generated, in addition to the normal one, without much trouble it is advisable to do so—particularly if the logic device under

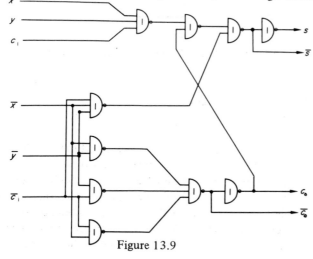

Figure 13.9

consideration is to be one of a chain of logic units. Should the negated inputs $\bar{x}, \bar{y}$, or $\bar{c}_i$ be not available as inputs to the circuit of Figure 13.9 then additional NOR units must be provided, to generate them from x, y, or c_i as necessary.

Subtractors

The unit which carries out the operation $x - y$ corresponding to the half-adder is known as a *half-subtractor*. This is also a two-input, two-output device and generates a *difference* bit, if only one of the inputs is a 1, and a *borrow* bit if

Table 13.6

	$x - y$			
y	0	1	0	1
x	0	0	1	1
Difference	0	1	1	0
Borrow	0	1	0	0

$x = 0$ and $y = 1$, otherwise both the difference and borrow bits are 0s. The truth table for this device is given in Table 13.6 and its block schematic in Figure 13.10.

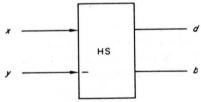

Figure 13.10

As in the case of the half-adder, the d (difference) is such that

$$d = x \not\equiv y$$

and, in a similar fashion, the b (borrow) is

$$b = \bar{x}y$$

Note that it is important, as far as b is concerned, to get the negation correct, i.e. $\bar{x}y$ refers to the subtraction of y from x and *not* x from y.

Since the logical expression for the difference of a half-subtractor is identical with that for the sum of a half-adder, the list of alternatives for the sum given in Table 13.2 can be used as the list for the difference; all that is necessary is to alter the heading of the table to difference $(x - y)$. The borrow term $\bar{x}y$ is *not* symmetrical in x and y and therefore the half-subtractor is not symmetrical in its arguments. This must be indicated as shown in Figure 13.10. Three alternative expressions for the borrow are listed in Table 13.7. The NOR versions of the circuits to generate d and b, corresponding to those which generate s and c of Figure 13.4(b), are shown in Figure 13.11(b).

Table 13.7

	Borrow
(a)	$\bar{x}y$
(b)	$\overline{(y \to x)}$
(c)	$(x + y)$

A *full-subtractor* must cater for the possibility of a borrow bit from a previous stage, just as a full-adder was required to handle a carry-in. Therefore, this is a three-input, two-output device in which the inputs are x, y, and b_i (the

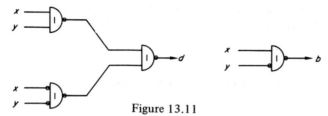

Figure 13.11

borrow-in from the previous stage) and the outputs are d and b_o (the *borrow-out* to the next stage). The block schematic of a full-subtractor is shown in Figure 13.12 and the description of its operation is as follows. A difference bit occurs if there are an *odd* number of input bits and a borrow bit occurs if

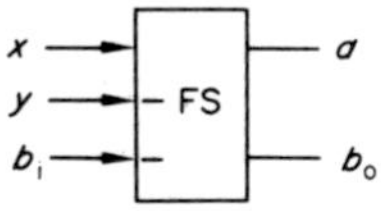

Table 13.8

y	0	1	0	1	0	1	0	1
x	0	0	1	1	0	0	1	1
b_i	0	0	0	0	1	1	1	1
d	0	1	1	0	1	0	0	1
b_o	0	1	0	0	1	1	0	1

Figure 13.12

$x = 0$ and either y or b_i or both equal 1 *or if all three inputs are* 1. Otherwise both d and b_o will both be 0. The relevant truth table is that of Table 13.8 and the basic formulae reduce to

$$d = x \not\equiv y \not\equiv b_i$$

and

$$b_o = y b_i + (y + b_i)\bar{x}$$

where the lack of symmetry in the expression for b_o can be clearly seen. An alternative pair of equations is

$$d = x \not\equiv y \not\equiv b_i$$

and

$$b_o = y b_i + (y \not\equiv b_i)\bar{x}$$

The form of this last pair of equations indicates how a full-subtractor can be formed from two *half-adders*. Although the above equations are not completely symmetric in x, y, and b_i, they *are* symmetric in y and b_i only, and so, if the inputs to a half-adder are made equal to y and b_i, its sum (s_1) will be $y \not\equiv b_i$ and its carry (c_1) will be $y b_i$. In terms of s_1 and c_1 the formulae for d and b_o become

$$d = x \not\equiv s_1$$

and

$$b_o = c_1 + s_1 \bar{x}$$

Obviously a second half-adder with inputs x and s_1 will generate d as its sum (s_2) but will only generate $s_1 x$ as its carry (c_2). However, b_o can be re-written, in terms of x, c_1, and c_2, as

$$b_o = (s_1 + c_1)\bar{c}_2$$

The validity of this can be demonstrated as follows. Since

$$c_2 = s_1 x$$

and

$$s_1 = y \not\equiv b_i$$

then

$$\bar{c}_2 = (y \equiv b_i) + \bar{x}$$
$$= y b_i + \bar{y}\bar{b}_i + \bar{x}$$

Further, since

$$c_1 = y b_i$$

and

$$s_1 = y \not\equiv b_i$$

then

$$(s_1 + c_1) = (y \not\equiv b_i) + y b_i$$
$$= y \bar{b}_i + \bar{y} b_i + y b_i$$
$$= y + b_i$$

Therefore,

$$(s_1 + c_1)\bar{c}_2 = (y + b_i)(y b_i + \bar{y}\bar{b}_i + \bar{x})$$
$$= y b_i + y\bar{x} + b_i\bar{x}$$
$$= y b_i + (y + b_i)\bar{x}$$
$$= b_o$$

Figure 13.13(a) shows the implementation of this approach. The use of a half-adder and a half-subtractor is shown in Figure 13.13(b) and that of two half-subtractors in Figure 13.13(c). These merely require the justification that

$$b_o = \bar{x}s + c_1$$

where

$$s_1 = y \not\equiv b_i$$

and

$$c_1 = y b_i$$

in the first case, and

$$b_o = b_2 + b_1$$

where

$$b_1 = \bar{x}y$$

and

$$b_2 = \bar{d}_1 b_i$$

where

$$d_1 = \bar{x}y$$

in the second.

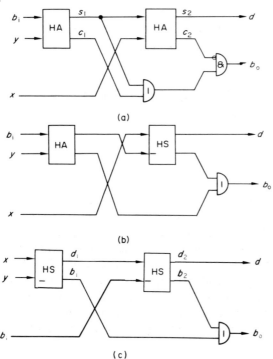

(a)

(b)

(c)

Figure 13.13

Considering the first case,

$$\bar{x}s_1 = \bar{x}(y\bar{b}_i + \bar{y}b_i)$$
$$= \bar{x}y\bar{b}_i + \bar{x}\bar{y}b_i$$

and

$$c_1 = xyb_i + \bar{x}yb_i$$

therefore

$$\bar{x}s_1 + c_1 = \bar{x}y\bar{b}_i + \bar{x}\bar{y}b_i + xyb_i + \bar{x}yb_i$$
$$= \bar{x}y + yb_i + \bar{x}b_i$$
$$= yb_i + (y + b_i)\bar{x}$$
$$= b_o$$

Turning now to the second case,

$$b_1 = \bar{x}yb_i + \bar{x}y\bar{b}_i$$

and

$$b_2 = (\overline{\bar{x}y})b_i$$
$$= (x + \bar{y})b_i$$
$$= xb_i + \bar{y}b_i$$
$$= xyb_i + x\bar{y}b_i + x\bar{y}b_i + \bar{x}\bar{y}b_i$$
$$= xyb_i + x\bar{y}b_i + \bar{x}\bar{y}b_i$$

therefore

$$b_1 + b_2 = \bar{x}yb_i + \bar{x}y\bar{b}_i + xyb_i + x\bar{y}b_i + \bar{x}\bar{y}b_i$$

which reduces to b_o as before.

Adder–Subtractors

In addition to signal levels corresponding to x and y, let there be a *control* signal p such that if $p = 1$ then addition, i.e. $x + y$,† must take place whilst if $p = 0$ then subtraction, i.e. $x - y$, is to be the operation implemented. The device to carry out this implementation is known as an *adder–subtractor*. The truth table shown in Table 13.9 describes the action of the *half-adder–subtractor*, where z is to be the sum and w the carry if $p = 1$, otherwise z is to be the difference and w the borrow. The block schematic of Figure 13.14 again requires some means of indicating, on the input lines, which input is to be added or subtracted as the case may be. Obviously, z must be independent of p and depend

† Bold plus sign (+) indicates arithmetical plus as apposed to 'or' which is signified by +.

only upon x and y, but w, on the other hand, does depend upon p. The Boolean equations defining z and w are

$$z = x \not\equiv y$$

and

$$w = \bar{x}y\bar{p} + xyp$$

i.e.

$$= y(x \equiv p)$$

Table 13.9

	$(x - y)$				$(x + y)$			
y	0	1	0	1	0	1	0	1
x	0	0	1	1	0	0	1	1
p	0	0	0	0	1	1	1	1
z	0	1	1	0	0	1	1	0
w	0	1	0	0	0	0	0	1

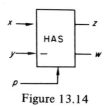

Figure 13.14

The NOR diagram of Figure 13.15 produces w, z, and their negations from x, y, p, and their negations.

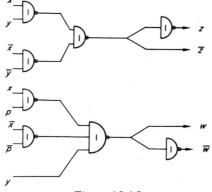

Figure 13.15

Full-Adder—Subtractors

The truth table describing a *full-adder—subtractor* is given in Table 13.10. The expressions for z and w_o reduce to

$$z = x \not\equiv y \not\equiv w_i$$

and

$$w_o = yw_i + (y \not\equiv w_i)(x \equiv p)$$

and the block schematic is shown in Figure 13.16. Note that the y and w_i inputs must again be indicated as being different from the x input, but no distinction need be made between them as z and w_o are symmetrical expressions in y and w_i.

Table 13.10

	(x − y)								(x + y)							
y	0	1	0	1	0	1	0	1	0	1	0	1	0	1	0	1
x	0	0	1	1	0	0	1	1	0	0	1	1	0	0	1	1
w_i	0	0	0	0	1	1	1	1	0	0	0	0	1	1	1	1
p	0	0	0	0	0	0	0	0	1	1	1	1	1	1	1	1
z	0	1	1	0	1	0	0	1	0	1	1	0	1	0	0	1
w_o	0	1	0	0	1	1	0	1	0	0	0	1	0	1	1	1

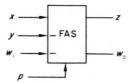

Figure 13.16

Complementors and Parallel Adders

An n-bit *parallel binary adder* is formed from a chain of n full-adders, with provision being made for carry from one section to the next. This is shown for $n = 4$ in Figure 13.17 where the least significant bits are at the right-hand side

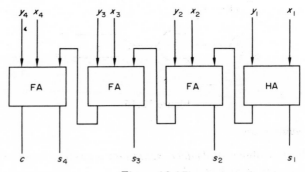

Figure 13.17

of the diagram. Note that only a half-adder is necessary for the first unit since only two of its inputs are used. Further, any carry from the last stage results in an overflow (OVR), thus indicating that the sum $x + y$ has exceeded the capacity of the device.

A *1's complementor* is a device used to complement a bit under logic control, i.e. it is distinguished from a simple negater by being switchable. If the control signal p is a 1 then the variable x is *not* negated, whereas if p is a 0 then x *is* negated. A logic diagram of such a device is shown in Figure 13.18. The principal

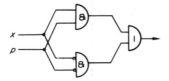

Figure 13.18

use of complementors is in sign and modulus arithmetic as described in Chapter 1.

Coded Decimal Adders

The most common coded decimal scale is the NBCD (natural binary coded decimal) and the addition of two decimal digits in this scale requires the addition of two four-bit binary numbers, perhaps using a device such as that of Figure 13.17. However, overflow will only occur with such a device if the resulting sum exceeds fifteen in magnitude and what is really required is that carry over into the next stage, representing the tens, must take place if the sum exceeds nine. The only valid representations allowed are the digits 0 to 9 inclusive. However, correct arithmetic will result, whenever the sum exceeds nine, if there is an indication that a 10's digit has been generated and the unit's sum is formed by subtracting ten from the original sum. This can obviously be achieved by subtracting sixteen and then *adding six*. Thus if overflow occurs, at the most significant end of the four-bit adder, then this overflow bit is sent to the 10's location and six is added to the remaining four bits by means of another four-bit adder. On the other hand, should overflow *not* take place, but the number representing the sum exceeds nine then six is added as before and the overflow which then takes place is again sent to the 10's location. These can obviously be combined into the following algorithm for addition in NBCD.

Form the sum of the two NBCD numbers using a four-bit adder. Test if this sum exceeds nine and, if it does, add six to it in another four-bit adder. If either adder overflows then transmit a bit to the next position (the 10's position).

The circuit for the testing for an invalid combination, i.e. a number which has exceeded nine, is derived from Table 13.11. In this table the sum bits s_i

have the weights 8, 4, 2, and 1 as usual and the overflow bit, represented by
c_4, has the weight 16. The decimal integers from 0 to 15 inclusive are capable

Table 13.11

	Bits	s_4	s_3	s_2	s_1	c_4
	Weights	8	4	2	1	16
Valid digits	0	0	0	0	0	0
	1	0	0	0	1	0
	2	0	0	1	0	0
	3	0	0	1	1	0
	4	0	1	0	0	0
	5	0	1	0	1	0
	6	0	1	1	0	0
	7	0	1	1	1	0
	8	1	0	0	0	0
	9	1	0	0	1	0
Invalid digits	10	1	0	1	0	0
	11	1	0	1	1	0
	12	1	1	0	0	0
	13	1	1	0	1	0
	14	1	1	1	0	0
	15	1	1	1	1	0
	16	0	0	0	0	1
	17	0	0	0	1	1
	18	0	0	1	0	1

of being represented by four bits, the only valid ones being 0 to 9 inclusive. An
invalid combination signal k is defined by

$$k = c_4 + s_4(s_3 + s_2)$$

because k is seen to be a 1 if c_4 is a 1 and also if s_4 is a 1 *and* s_3 and s_2 are not
both 0 together. The circuit which produces this is shown in Figure 13.19 and

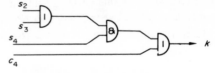

Figure 13.19

an NBCD adder unit employing seven full-adders is shown in Figure 13.20. This latter circuit uses the k as generated by Figure 13.19, the final NBCD number is given by s'_1, s'_2, s'_3, and s'_4 and the ten's digit is indicated by a 1 on the t line.

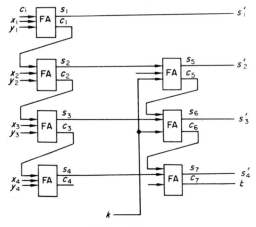

Figure 13.20

Code Converters

In this section, it will frequently be useful to indicate the decimal digits 0 to 9 inclusive by d_0 to d_9 inclusive, i.e. as if the ten-bit odd parity code of Table 13.12, having weights 0 to 9 inclusive, were to be adopted. One application of code converters is in the linking of two pieces of equipment of different manufacture and using different numerical representation. The justification for their inclusion here, however, is in their use in arithmetical units. An adder can be produced

Table 13.12

| | | Weights | | | | | | | | | |
		9	8	7	6	5	4	3	2	1	0
	d_0	0	0	0	0	0	0	0	0	0	1
	d_1	0	0	0	0	0	0	0	0	1	0
	d_2	0	0	0	0	0	0	0	1	0	0
Valid	d_3	0	0	0	0	0	0	1	0	0	0
decimal	d_4	0	0	0	0	0	1	0	0	0	0
digits	d_5	0	0	0	0	1	0	0	0	0	0
	d_6	0	0	0	1	0	0	0	0	0	0
	d_7	0	0	1	0	0	0	0	0	0	0
	d_8	0	1	0	0	0	0	0	0	0	0
	d_9	1	0	0	0	0	0	0	0	0	0

based upon *any* coded decimal representation—by techniques similar to those used in the previous sections of this chapter. However, it is possible to take such a coded decimal representation and convert it to another representation by logic circuitry. The arithmetic is then performed using the circuits already designed. The result can then be converted back into the original code.

Decimal to NBCD

The logic equations for the conversion of a decimal to an NBCD number are

$$s_4 = d_8 + d_9$$
$$s_3 = d_4 + d_5 + d_6 + d_7$$
$$s_2 = d_2 + d_3 + d_6 + d_7$$

and

$$s_1 = d_1 + d_3 + d_5 + d_7 + d_9$$

obtained by comparing the valid digits of Table 13.11 and 13.12.

The reverse procedure is given by

$$d_9 = s_1 + s_4$$
$$d_8 = \bar{s}_1 + s_4$$
$$d_7 = s_1 + s_2 + s_3$$
$$d_6 = \bar{s}_1 + s_2 + s_3$$
$$d_5 = s_1 + \bar{s}_2 + s_3$$
$$d_4 = \bar{s}_1 + \bar{s}_2 + s_3$$
$$d_3 = s_1 + s_2 + \bar{s}_3$$
$$d_2 = \bar{s}_1 + s_2 + \bar{s}_3$$
$$d_1 = s_1 + \bar{s}_2 + \bar{s}_3 + s_4$$

and

$$d_0 = \bar{s}_1 + \bar{s}_2 + \bar{s}_3 + \bar{s}_4$$

The latter are easier to justify if they are considered in three sets $-\{d_0, d_1\}$ followed by $\{d_2, d_3, d_4, d_5, d_6, d_7\}$, and $\{d_8, d_9\}$ and noting the dependence of the elements of these sets upon the bit positions of Table 13.11.

Johnson Code to NBCD

In this instance, an unweighted code (the Johnson code of Table 13.13) is converted to a weighted code (NBCD). Two of the relationships are obvious, namely

$$s_4 = \bar{j}_3 j_5$$
$$s_3 = j_3 j_4$$

and the third is seen to be

$$s_2 = j_1 j_2 \bar{j}_4 \bar{j}_5 + \bar{j}_1 j_3 j_4 j_5$$

which, in view of the large number of don't care terms, corresponding to the invalid combinations, simplifies to

$$s_2 = j_2 \bar{j}_4 + \bar{j}_1 j_3$$

Table 13.13

Johnson					NBCD			
j_5	j_4	j_3	j_2	j_1	s_4	s_3	s_2	s_1
0	0	0	0	0	0	0	0	0
0	0	0	0	1	0	0	0	1
0	0	0	1	1	0	0	1	0
0	0	1	1	1	0	0	1	1
0	1	1	1	1	0	1	0	0
1	1	1	1	1	0	1	0	1
1	1	1	1	0	0	1	1	0
1	1	1	0	0	0	1	1	1
1	1	0	0	0	1	0	0	0
1	0	0	0	0	1	0	0	1

The final equation can be written as

$$s_1 = j_1 \bar{j}_2 \bar{j}_3 \bar{j}_4 \bar{j}_5 + j_1 j_2 j_3 \bar{j}_4 \bar{j}_5 + j_1 j_2 j_3 j_4 j_5 + \bar{j}_1 \bar{j}_2 j_3 j_4 j_5 + \bar{j}_1 \bar{j}_2 \bar{j}_3 j_4 j_5$$

which simplifies to

$$s_1 = j_1 \bar{j}_2 + j_1 j_5 + j_3 \bar{j}_4 + \bar{j}_4 j_5 + \bar{j}_2 j_3$$

Conversely

$$j_1 = (s_1 \bar{s}_2 + s_2 \bar{s}_3 + \bar{s}_2 s_3) \bar{s}_4$$
$$j_2 = (\bar{s}_1 s_2 + s_2 \bar{s}_3 + \bar{s}_2 s_3) \bar{s}_4$$
$$j_3 = (s_3 + s_1 s_2) \bar{s}_4$$
$$j_4 = s_3 \bar{s}_4 + \bar{s}_1 s_4$$
$$j_5 = s_4 + (s_1 + s_2) s_3$$

The NOR circuit to realize the last five equations is given in Figure 13.21.

Multipliers

With the introduction of relatively cheap integrated circuits, it became economically feasible to construct parallel hardware multipliers. The principles,

involving simple combinations of logic gates and binary adders, have been
known for some time; only the cost has been prohibitive. In these multipliers,
the bits of the multiplier and the multiplicand are applied simultaneously and,
after the transients have died down, the product is available at the output
terminals, as long as the inputs remain active.

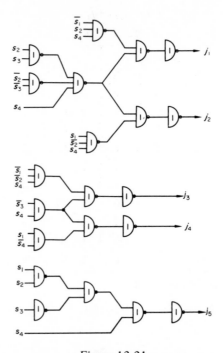

Figure 13.21

One method of realizing this process of multiplication is analogous to direct
multiplication by hand, i.e. the partial products are first formed, as in long
multiplication, and these are then summed.

Example. Form the product of the two binary numbers 1 0 1 1 and
1 0 0 1.

				1	0	1	1	multiplicand
				1	0	0	1	multiplier
				1	0	1	1	1st partial product
			0	0	0	0		2nd partial product
		0	0	0	0			3rd partial product
	1	0	1	1				4th partial product
0	1	1	0	0	0	1	1	Product

Using a suffix notation with three-bit numbers, the scheme would look like

$$
\begin{array}{ccc}
x_3 & x_2 & x_1 \\
y_3 & y_2 & y_1 \\
\hline
x_3 y_1 & x_2 y_1 & x_1 y_1 \\
x_3 y_2 & x_2 y_2 & x_1 y_2 \\
x_3 y_3 & x_2 y_3 & x_1 y_3 \\
\end{array}
$$

$$
\begin{array}{cccccc}
z_6 & z_5 & z_4 & z_3 & z_2 & z_1
\end{array}
$$

The extension of this scheme to one involving a larger number of bits is obvious.

The logic diagram makes use of the fact that the partial products are the result of using the conjunction operation since all the x_i and y_i can only be

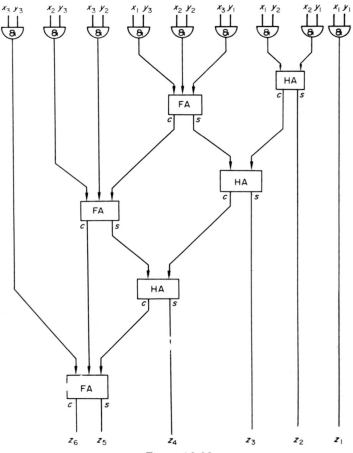

Figure 13.22

236

LOGIC AND LOGIC DESIGN

either 0 or 1. There is no question of a carry from a partial product into the next position; this comes from the summation of the partial products at the end.

In the multipliers to be considered, $z_1 = x_1 y_1$ and, since no other partial product is involved, a simple AND gate will suffice. The 3 x 3 multiplication scheme shown above is realized by nine AND gates for the partial products, a half-adder for z_2, a half-adder and a full-adder for z_3 and also for z_4, whilst z_5 and z_6 require a full-adder. Hence, for the three-bit multiplier, a total of nine AND gates, three half-adders, and three full-adders are required to form the six outputs. This scheme is shown in Figure 13.22.

Figure 13.23 shows the extension to a 4 x 4 multiplier, and it can be seen to involve sixteen AND gates, four half-adders, and eight full-adders.

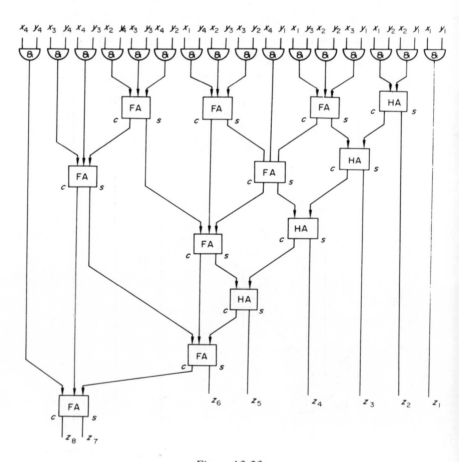

Figure 13.23

The number of partial products which have to be summed, to form the components of the final product z_i is given in Table 13.14 for an $m \times m$ multiplier. Note that z_1 is the least significant bit and z_{2m-1} the most significant.

Table 13.14

i	$1 \quad 2 \quad 3 \quad \ldots \quad (m-1)\, m\, (m+1) \quad \ldots \quad (2m-1)$
Number of partial products in the summation of z_i	$1 \quad 2 \quad 3 \quad \ldots \quad (m-1)\, m\, (m-1) \quad \ldots \quad 3 \quad 2 \quad 1$

The systematic determination of the number of half- and full-adders is laid out in Table 13.15, for the case when $m = 5$. The integers in this table denote the number of inputs that are to be fed into an adder, with the exception of the least significant one which will always be the output of a simple two-input

Table 13.15

i	1	2	3	4	5	6	7	8	9
Partial products	1	2	3	4	5	4	3	2	1
									&
							4	H	
						5	F, 1		
						6	H		
					6	F, 3			
					7	F, 1			
					8	H			
				5	F, 5				
				6	F, 3				
				7	F, 1				
				8	H				
			4	F, 5					
			5	F, 3					
			6	F, 1					
			7	H					
		3	F, 4						
		4	F, 2						
		5	F, 0						
	2	F, 2							
	3	F, 0							
	F, 0								

AND gate. In any column, an integer k followed by a $k + 1$ will indicate that a full- or half-adder to the right of the $k + 1$ has caused an increase of unity in the number of inputs for that column. An integer k followed by F, $k - 3$ indicates that a full-adder has taken care of three of the inputs, hence the reduction to $k - 3$; at the same time the carry from the adder will increase by

unity the number of inputs in the adjacent column to the left. Note that an
F, $k - 3$ will be followed by F, $k - 5$; F, $k - 7$, etc., until either F, 0 or F, 1 is
reached. F, 0 means that the last element is a full-adder. F, 1 will *always* be
followed by H, i.e. a half-adder, with a corresponding increase by unity in the
number of inputs in the adjacent column to the left.

Table 13.16

n	&	H	F
1	1	0	0
2	4	2	0
3	9	3	3
4	16	4	8
5	25	5	15
6	36	6	24

Table 13.16 lists the number of AND gates, half-adders and full-adders
required for an $n \times n$ multiplier unit of the above design.

Suggestions for Further Reading

Baron, R. C., and Piccinilli, A. T., 1967, *Digital Logic and Computer
  Applications*, McGraw-Hill, New York.
Caldwell, S. H., 1968, *Switching Circuits and Logic Design*, Wiley, New York.
Chu, Y., 1962, *Digital Computer Design Fundamentals*, McGraw-Hill, New York.
Wickles, W. E., 1968, *Logic Design with Integrated Circuits*, Wiley, New York.

Exercises

1. Design three full-subtractors, using NAND elements only, based upon the
   block diagrams of Figure 13.13.

2. Using seven full-adders, design an XS3 adder unit corresponding to the
   NBCD adder unit of Figure 13.20.

3. Design three code converters:
   (i) to convert from XS3 to NBCD
   (ii) to convert from NBCD to XS3
   (iii) to convert from Johnson code to XS3
   In all three cases, the only valid combinations are to be those corresponding to the decimal digits 0 to 9 inclusive.

4. Design a logic unit which will form the 9's complement, using NBCD, of a four-bit number on the receipt of a 'complement' signal.

5. Draw up the table for converting from decimal to a code having the weights 5, 4, 2, and 1 and design an adder unit, using this code, similar to that of Figure 13.20.

6. A four-bit NBCD number is to be transformed into a five-bit number. The additional bit is intended to be a parity checking bit. Assuming that the parity mode is to be that of odd parity, design the logic circuit which will generate this from the original NBCD input.

7. Justify the values given in Table 13.16.

Chapter 14
Sequential Circuits and Constraints

Sequential Circuits

An example of such a circuit is given by a *recursive* circuit, in which some or all of the outputs are fed back through delay units to form additional inputs. This is not strictly Boolean logic since time is now an integral part of the circuit behaviour pattern and use is made of *timing diagrams* to illustrate the behaviour of the circuits under consideration. These are diagrams in which time is read from left to right, i.e. a pulse shown on the left of a timing diagram is assumed to have existed before another pulse shown on the right of the same diagram. Figure 14.1 shows the timing diagram for a set of *clock* pulses a, b, c, d, . . ., etc., using the positive logic convention. These are a set of rectangular pulses all having the same pulse width w and the same interval Δ between the starts or leading edges of successive pulses. In general, they will all be of the same height but this has no particular bearing upon the logical behaviour. In practice it is usual for them to have to exceed a certain minimum for the component to work. Returning to Figure 14.1, if the pulse width w is 1 μs (say) and the pulse separation 'Δ' is 4 μs, then, if pulse 'a' starts at a time $t = 0$, pulse 'b' will start

Figure 14.1

at $t = 4$ μs, 'c' at $t = 8$ μs, etc., each pulse lasting for 1 μs. It should be noted that logic signals will be significant only when a clock pulse is operating—this is known as *clocked* or *synchronous* logic. It is illustrated in Figure 14.2 where the signals x, y, and z will have the same effect logically upon any circuit for which they are the inputs. The cases where logical signals are *not* synchronized with clock pulses (asynchronous logic) will not be considered here.

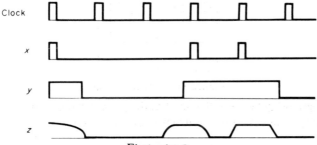

Figure 14.2

A *sequential* circuit is a logic circuit that produces an output sequence of 0s and 1s in time which is a function of the previous history of the circuit and possibly one or more input variables.

Delays

The symbol for a *unit delay*, i.e. a delay of one clock period, will consist of the symbol 'Δ' placed in a break in the appropriate connecting line of the logic diagram. A delay of n clock periods will have the symbol 'Δ^n' placed in the break. A delay of two clock periods is shown diagrammatically in Figure 14.3,

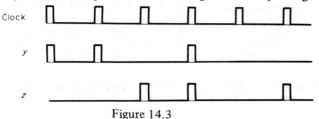

Figure 14.3

where the output z is simply the input y delayed by two clock periods. In order to avoid drawing complete timing diagrams, they will be represented by a sequence of 0s and 1s, where a 1 indicates the presence of a pulse and a 0 indicates the absence. For example, Figure 14.3 can be replaced by

clock	1	1	1	1	1	1
y	1	1	0	1	0	0
z	0	0	1	1	0	1

Such sequences will be read from left to right, i.e. the left-most digit corresponds to the first pulse in time. Two unit delays are incorporated in the simple recursive circuit of Figure 14.4. In this figure, suppose that the logic is such

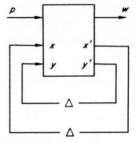

Figure 14.4 Mealy model

that the relationship between the input p, the present states x and y, the next states x' and y' and the output w are given by Table 14.1, i.e.

$$x' = py + \bar{p}x$$
$$y' = p\bar{x} + \bar{p}y$$

and

$$w = p(x \not\equiv y) + \bar{p}xy$$

Table 14.1

Present state	y	0	1	0	1	0	1	0	1
	x	0	0	1	1	0	0	1	1
Input	p	0	0	0	0	1	1	1	1
Next state	y'	0	1	0	1	1	1	0	0
	x'	0	0	1	1	0	1	0	1
Output	w	0	0	0	1	0	1	1	0

Note that a next state and its corresponding present state are *always* connected via a delay.

Mealy Models

The recursive circuit of Figure 14.4 is an example of a *Mealy* model of a sequential circuit (Booth, 1967). In this type of model, as is shown, the outputs and the *next states* are functions of the inputs and the *present states*. However, in practice, some or all of the outputs may be identical with some or all of the next states. For example, if in Table 14.1 the output w had been represented by the designation number 0 0 1 1 0 1 0 1, with respect to the basis

$b[y, x, p]$, instead of 0 0 0 1 0 1 1 0 as shown, then w and x' would have been identical and the output could have been taken from the x' line.

Moore Models

If, in a recursive circuit, the outputs are logic functions of the present states only, i.e. they are functions which do not directly include the inputs, then the circuit constitutes a *Moore* model (Miller, 1965) of a sequential circuit. Figure 14.5 shows a typical example of such a model. The figure is drawn as shown to emphasize the point that the only contribution of the input to the output is via the present state variables.

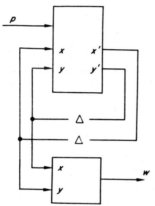

Figure 14.5 Moore model

For example, if the relationship between the variables shown in Figure 14.5 is as tabulated in Table 14.2, it can be seen that the outputs depend only upon the present states.

Table 14.2

Present state	y	0	1	0	1	0	1	0	1
	x	0	0	1	1	0	0	1	1
Input	p	0	0	0	0	1	1	1	1
Next state	y'	0	1	0	1	1	1	0	0
	x'	0	0	1	1	0	1	0	1
Output	w	0	1	1	0	0	1	1	0

State Tables

Tables 14.1 and 14.2 are known as *binary state tables* and completely specify the sequential circuits with which they are associated. However, if there are more than one state, or input, or output, then it is usual to attach the weights

1, 2, 4, etc., to the various members of the set of states and hence tabulate their decimal equivalents. Tables 14.3 and 14.4 are thus the *decimal state tables* which are equivalent to the binary state tables previously mentioned. Given Table 14.5 as the decimal state table of a sequential machine, it can be seen by inspection

Table 14.3

Present state	0	1	2	3	0	1	2	3
Input	0	0	0	0	1	1	1	1
Next state	0	1	2	3	1	3	0	2
Output	0	0	0	1	0	1	1	0

Table 14.4

Present state	0	1	2	3	0	1	2	3
Input	0	0	0	0	1	1	1	1
Next state	0	1	2	3	1	3	0	2
Output	0	1	1	0	0	1	1	0

Table 14.5

Present state	0	1	2	3	0	1	2	3	0	1	2	3	0	1	2	3
Input	0	0	0	0	1	1	1	1	2	2	2	2	3	3	3	3
Next state	1	2	2	3	0	1	0	3	2	2	1	3	3	1	0	0
Output	0	1	1	0	1	1	0	1	0	1	0	1	0	0	0	1

that there are two state variables and two inputs but only one output. Moreover, the model in question is a Mealy model as the output values depend upon the inputs. Consider now the Mealy model of Figure 14.6(a), this has a unit delay

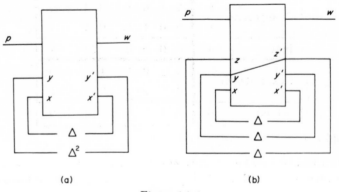

(a)   (b)

Figure 14.6

and a delay of 2 incorporated in the circuit. This can be transformed by the simple device shown in Figure 14.6(b), i.e. the delay of 2 can be split into two unit delays—provided a third state variable is introduced and one of the present state variables is renamed.

Table 14.6

Present state	0 1 2 3	0 1 2 3
Input	0 0 0 0	1 1 1 1
Next state	2 2 0 1	3 0 1 2
Output	0 1 1 0	0 0 1 1

In Figure 14.6(b), the present state z corresponds to the present state y of Figure 14.6(a) the present state x and the next states x' and y' remaining unaltered. Because of this transposition, it is advisable to use the binary state table representation at this stage. Consider again Figure 14.6(a), and let the decimal state table associated with this figure be that of Table 14.6. To find the state table, associated with the corresponding Figure 14.6(b), proceed as follows.

Table 14.7

Present state	y	0 1 0 1	0 1 0 1
	x	0 0 1 1	0 0 1 1
Input	p	0 0 0 0	1 1 1 1
Next state	y'	0 0 0 1	1 0 1 0
	x'	1 1 0 0	1 0 0 1
Output	w	0 1 1 0	0 0 1 1

(a)

y	0 1 0 1	0 1 0 1	0 1 0 1	0 1 0 1
q	0 0 0 0	0 0 0 0	1 1 1 1	1 1 1 1
x	0 0 1 1	0 0 1 1	0 0 1 1	0 0 1 1
p	0 0 0 0	1 1 1 1	0 0 0 0	1 1 1 1
z'	0 1 0 1	0 1 0 1	0 1 0 1	0 1 0 1
y'	0 0 0 1	1 0 1 0	0 0 0 1	1 0 1 0
x'	1 1 0 0	1 0 0 1	1 1 0 0	1 0 0 1
w	0 1 1 0	0 0 1 1	0 1 1 0	0 0 1 1

(b)

Replace the decimal present states of Table 14.6 by their binary equivalents as in Table 14.7(a). Take the first three rows of Table 14.7(a) and treat them as the basis $b[y, x, p]$. Extend this to form the new basis $b[y, x, p, q]$ in the usual manner. Duplicate the rows, corresponding to the next states and outputs, one group being associated with $q = 0$ and the other with $q = 1$. Insert a row labelled z', every truth value of which is equal to the truth value of the row labelled y above it. The rows labelled x, p, and q are then permuted cyclically, i.e. row p is replaced by row x, row q by row p, and row x by row q, while the rows labelled y, z', y', x', and w remain *in situ*. The result of these movements is shown in Table 14.7(b).

The columns of Table 14.7(b) are then rearranged to produce the usual basis lay-out. Finally the row labelled y is relabelled z and the row labelled q is relabelled y. This is the basis $b[z, y, x, p]$ of Table 14.8.

Table 14.8

z	0	1	0	1	0	1	0	1	0	1	0	1	0	1	0	1
y	0	0	1	1	0	0	1	1	0	0	1	1	0	0	1	1
x	0	0	0	0	1	1	1	1	0	0	0	0	1	1	1	1
p	0	0	0	0	0	0	0	0	1	1	1	1	1	1	1	1
z'	0	0	1	1	0	0	1	1	0	0	1	1	0	0	1	1
y'	0	0	0	0	0	1	0	1	1	0	1	0	1	0	1	0
x'	1	1	1	1	0	0	0	0	1	0	1	0	0	1	0	1
w	0	1	0	1	1	0	1	0	0	0	0	0	1	1	1	1

Conversion of General Sequential Circuits to Mealy Models

Most logic circuits involving delays can be redrawn in the form of a Mealy model. For example, Figure 14.7(a) can be redrawn in the form of Figure 14.7(b) without altering the input—output—state relationships. The conversion to a Moore model representation will not always be possible since this model restricts the output to be a function of the present states only. However, Figure 14.8 illustrates a case in which this conversion does work.

Transition Tables

A *transition table*, sometimes referred to as a *flow table* or *Huffman table* is shown in Table 14.9. In this type of table, the columns are labelled with the present inputs and the rows with the present states. The entries in the table consist of the corresponding next states and present outputs, separated by a '/'. Should any entry be undefined or indeterminate then it is represented by a

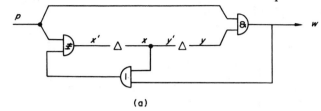

(a)

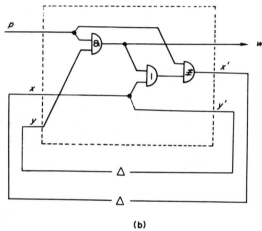

(b)

Figure 14.7 Conversion to a Mealy model

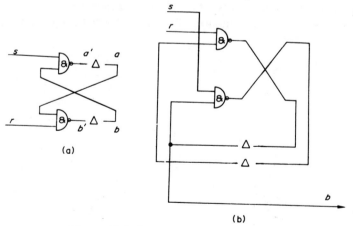

Figure 14.8 Conversion to a Moore model

'—'. Transition tables are merely transformed state tables and both the binary and decimal forms may be employed. For example, Table 14.1 can be expressed in the form of the transition table of Table 14.10 and Table 14.5 in the form of Table 14.11.

Table 14.9

		Present input	
		0	1
Present state	0	2/0	3/0
	1	2/1	—/0
	2	0/0	—/—
	3	3/1	4/1
	4	1/—	3/1

Table 14.10

		Present input	
		0	1
Present state	00	00/0	01/0
	01	01/0	11/1
	10	10/0	00/1
	11	11/1	11/0

Table 14.11

		Present input			
		0	1	2	3
Present state	0	1/0	0/1	2/0	3/0
	1	2/1	1/1	2/1	1/0
	2	2/1	0/0	1/0	0/0
	3	3/0	3/1	3/1	0/1

State Diagrams

Yet another alternative representation, of the relationship between the inputs, the outputs, and the states, exists. This is the *state diagram*. For the Mealy model, the state diagram consists of a number of circles enclosing the states and

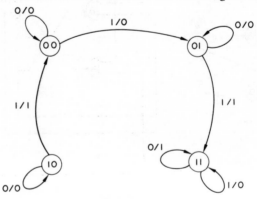

Figure 14.9

connected by directed lines indicating the present input which brings about the change of state and the present output. The present input and the present output are separated by a '/' the input preceding the output. Figure 14.9 is the state diagram corresponding to Table 14.10.

When a number of directed lines start and finish on the same states, as in Figure 14.10, it is only necessary to draw one line and attach all the input— output groups separated by commas. Figure 14.11 shows the simplification

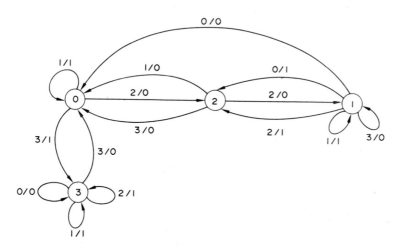

Figure 14.10

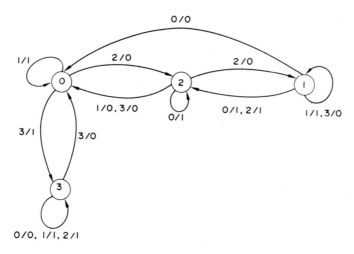

Figure 14.11

that results when this technique is applied to Figure 14.10. If an output is undefined then it is represented by a '—' on the state diagram, whereas if the next state is undefined the directed line will have a specific state as its origin but will be lacking a terminating state. Figure 14.12 corresponds to Table 14.9 and illustrates these points.

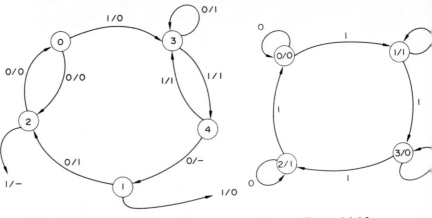

Figure 14.12   Figure 14.13

In the state diagram associated with the Moore model, the state and present outputs are both placed inside the circle and separated by a '/'; the only quantity associated with the directed line connecting the state/outputs being the input in question. Table 14.4, in which the outputs are a function of the present states only, can be taken as the state table of a Moore model. The corresponding state diagram is that of Figure 14.13. Note that the transition table for a Moore model will have the same output value for all the entries in any one row. Table 14.12 is the transition table relating to Figure 14.13.

Table 14.12

		Present input	
		0	1
	0	0/0	1/0
Present	1	1/1	3/1
state	2	2/1	0/1
	3	3/0	2/0

Transition Matrices

A *transition matrix* $\mathbf{T}^p$ is associated with each input 'p', the elements of $\mathbf{T}^p$ being denoted by $[t_{ij}^p]$.

There is one row and one column for each state such that $t_{ij}^p = 1$ if input p takes state i into state j, $t_{ij}^p = 0$ if input p takes state i into state m $(\neq j)$, and $t_{ij}^p = -$ (dash) otherwise.

A dash as an entry in the matrix indicates that the application of an input p to a sequential machine in a state i results in an undefined state. For example, the Moore model of Figure 14.13 has as its transition matrices

$$
\mathbf{T^0} = \begin{bmatrix} 1 & 0 & 0 & 0 \\ 0 & 1 & 0 & 0 \\ 0 & 0 & 1 & 0 \\ 0 & 0 & 0 & 1 \end{bmatrix}
\quad \text{and} \quad
\mathbf{T^1} = \begin{bmatrix} 0 & 1 & 0 & 0 \\ 0 & 0 & 0 & 1 \\ 1 & 0 & 0 & 0 \\ 0 & 0 & 1 & 0 \end{bmatrix}
$$

State and Output Vectors for Moore Models

A column vector q known as the *state vector* is formed from the ordered set of states and another column vector w known as the *output vector* is formed from the outputs associated with the corresponding states. For example, the Moore model of Figure 14.13 yields the state and output vectors

$$
q = \begin{bmatrix} 0 \\ 1 \\ 2 \\ 3 \end{bmatrix}
\quad \text{and} \quad
w = \begin{bmatrix} 0 \\ 1 \\ 1 \\ 0 \end{bmatrix}
$$

Note that the positions of the elements of q could be rearranged, but if this is done then the elements of w must suffer the same rearrangement.

As with the transition matrices, if any state q_i $(q = \{q_i\})$ or any output $w_j (w = \{w_j\})$ is undefined or indeterminate then the element in the appropriate position of the column vector is represented by a dash.

State and Output Vectors for Mealy Models

The definition of the state vector is the same as for the Moore model but, since the outputs depend upon the inputs, the number of output vectors equals the number of inputs. Hence, if w_{jp} is the output associated with the input p and the present state q_j then $w = [w_{jp}]$. As before, if any output is undefined it is represented by a dash. The transition matrices, the state vector and the output vectors for the Mealy model of Figure 14.12 are

$$
T^0 = \begin{bmatrix} 0 & 0 & 1 & 0 & 0 \\ 0 & 0 & 1 & 0 & 0 \\ 1 & 0 & 0 & 0 & 0 \\ 0 & 0 & 0 & 1 & 0 \\ 0 & 1 & 0 & 0 & 0 \end{bmatrix}, \quad
T^1 = \begin{bmatrix} 0 & 0 & 0 & 1 & 0 \\ - & - & - & - & - \\ - & - & - & - & - \\ 0 & 0 & 0 & 0 & 1 \\ 0 & 0 & 0 & 1 & 0 \end{bmatrix}
$$

$$
q = \begin{bmatrix} 0 \\ 1 \\ 2 \\ 3 \\ 4 \end{bmatrix}, \quad
w^0 = \begin{bmatrix} 0 \\ 1 \\ 0 \\ 1 \\ - \end{bmatrix} \quad \text{and} \quad
w^1 = \begin{bmatrix} 0 \\ 0 \\ - \\ 1 \\ 1 \end{bmatrix}
$$

It will frequently be convenient to replace the decimal values of the states by their binary equivalents. In the above example the state vector $\{0, 1, 2, 3, 4\}$ would be replaced by $\{000, 001, 010, 011, 100\}$.

Note the following.

(i) If any row of T^p contains *one* blank, then all the elements of that row are blank.

(ii) Every non-blank row of T^p contains one and only one 1.

(iii) The operations involved in the logical matrix product of Chapter 11 are augmented to deal with the case of blank entries as follows:

$$
\begin{aligned}
- \cdot 0 &= 0 \\
- \cdot 1 &= - \\
- \cdot - &= - \\
- + 0 &= - \\
- + 1 &= 1 \\
- + - &= -
\end{aligned}
$$

State Sequences and Output Sequences

If a sequential machine is in a state q_k and an input p is applied, then the next state is given by the kth row of $T^p \times q$ and, for a Moore model, the next output is given by the kth row of $T^p \times w$. The next state of a Mealy model is found in exactly the same fashion but, because of the multiplicity of output vectors, the determination of the next output is more complex—it will be discussed later in this chapter.

For example, using the Moore model represented by Figure 14.13

$$\mathbf{T}^0 \times q = \begin{bmatrix} 1 & 0 & 0 & 0 \\ 0 & 1 & 0 & 0 \\ 0 & 0 & 1 & 0 \\ 0 & 0 & 0 & 1 \end{bmatrix} \times \begin{bmatrix} 0 \\ 1 \\ 2 \\ 3 \end{bmatrix}$$

$$= \begin{bmatrix} 1 & 0 & 0 & 0 \\ 0 & 1 & 0 & 0 \\ 0 & 0 & 1 & 0 \\ 0 & 0 & 0 & 1 \end{bmatrix} \times \begin{bmatrix} 00 \\ 01 \\ 10 \\ 11 \end{bmatrix}$$

$$= \begin{bmatrix} 00 \\ 01 \\ 10 \\ 11 \end{bmatrix}$$

$$= \begin{bmatrix} 0 \\ 1 \\ 2 \\ 3 \end{bmatrix}$$

$$\mathbf{T}^1 \times q = \begin{bmatrix} 0 & 1 & 0 & 0 \\ 0 & 0 & 0 & 1 \\ 1 & 0 & 0 & 0 \\ 0 & 0 & 1 & 0 \end{bmatrix} \times \begin{bmatrix} 00 \\ 01 \\ 10 \\ 11 \end{bmatrix}$$

$$= \begin{bmatrix} 01 \\ 11 \\ 00 \\ 10 \end{bmatrix}$$

$$= \begin{bmatrix} 1 \\ 3 \\ 0 \\ 2 \end{bmatrix}$$

i.e. that state vector $\{0, 1, 2, 3\}$ becomes $\{0, 1, 2, 3\}$ after an input of 0 and $\{1, 3, 0, 2\}$ after an input of 1.

Example. Consider the Moore model having the state diagram of Figure 14.14(a). If the input sequence 0, 1, 1, 1 is presented to the model, determine, with respect to the initial state vector $\{1, 2, 3\}$, (a) the final state vector, and (b) the final output vector.

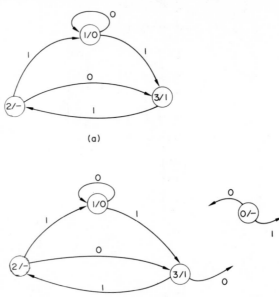

(a)

(b)

Figure 14.14

Two points are worthy of note. The first is that the state 0 does not exist on the state diagram and the second is that no provision is made for an input of 0 when the model is in the state 3.

Figure 14.14(b) indicates the modifications to the diagram which will involve all the possible states and inputs. These modifications are not normally inserted since Table 14.13, the transition table, can be produced directly from Figure 14.14(a).

Table 14.13

		Present input	
		0	1
	0	$-/-$	$-/-$
Present	1	1/0	3/1
state	2	3/1	3/0
	3	$-/-$	2/$-$

The transition matrices $\mathbf{T}^0$ and $\mathbf{T}^1$ are

$$\mathbf{T}^0 = \begin{bmatrix} - & - & - & - \\ 0 & 1 & 0 & 0 \\ 0 & 0 & 0 & 1 \\ - & - & - & - \end{bmatrix} \quad \text{and} \quad \mathbf{T}^1 = \begin{bmatrix} - & - & - & - \\ 0 & 0 & 0 & 1 \\ 0 & 1 & 0 & 0 \\ 0 & 0 & 1 & 0 \end{bmatrix}$$

and the binary form of the state vector is $\{00, 01, 10, 11\}$. The first input yields $\mathbf{T}^0 \times \boldsymbol{q}$, the second $\mathbf{T}^1 \times \mathbf{T}^0 \times \boldsymbol{q}$, the third $\mathbf{T}^1 \times \mathbf{T}^1 \times \mathbf{T}^0 \times \boldsymbol{q}$ and finally the fourth yields $\mathbf{T}^1 \times \mathbf{T}^1 \times \mathbf{T}^1 \times \mathbf{T}^0 \times \boldsymbol{q}$. In evaluating this logical matrix product the $\mathbf{T}$'s are first multiplied out and then the resulting square matrix is used to pre-multiply the state vector and the output vector. In this case

$$\mathbf{T}^1 \times \mathbf{T}^1 \times \mathbf{T}^1 \times \mathbf{T}^0 = \begin{bmatrix} - & - & - & - \\ 0 & 0 & 0 & 1 \\ 0 & 1 & 0 & 0 \\ 0 & 0 & 1 & 0 \end{bmatrix} \times \begin{bmatrix} - & - & - & - \\ 0 & 0 & 0 & 1 \\ 0 & 1 & 0 & 0 \\ 0 & 0 & 1 & 0 \end{bmatrix} \times \begin{bmatrix} - & - & - & - \\ 0 & 0 & 0 & 1 \\ 0 & 1 & 0 & 0 \\ 0 & 0 & 1 & 0 \end{bmatrix} \times \begin{bmatrix} - & - & - & - \\ 0 & 1 & 0 & 0 \\ 0 & 0 & 0 & 1 \\ - & - & - & - \end{bmatrix}$$

$$= \begin{bmatrix} - & - & - & - \\ 0 & 0 & 1 & 0 \\ 0 & 0 & 0 & 1 \\ 0 & 1 & 0 & 0 \end{bmatrix} \times \begin{bmatrix} - & - & - & - \\ 0 & 0 & 0 & 1 \\ 0 & 1 & 0 & 0 \\ 0 & 0 & 1 & 0 \end{bmatrix} \times \begin{bmatrix} - & - & - & - \\ 0 & 1 & 0 & 0 \\ 0 & 0 & 0 & 1 \\ - & - & - & - \end{bmatrix}$$

$$= \begin{bmatrix} - & - & - & - \\ 0 & 1 & 0 & 0 \\ 0 & 0 & 1 & 0 \\ 0 & 0 & 0 & 1 \end{bmatrix} \times \begin{bmatrix} - & - & - & - \\ 0 & 1 & 0 & 0 \\ 0 & 0 & 0 & 1 \\ - & - & - & - \end{bmatrix}$$

$$= \begin{bmatrix} - & - & - & - \\ 0 & 1 & 0 & 0 \\ 0 & 0 & 0 & 1 \\ - & - & - & - \end{bmatrix}$$

$$= \mathbf{R} \text{ (say)}$$

the final state vector being given by $\mathbf{R} \times \boldsymbol{q}$, where

$$\mathbf{R} \times q = \begin{bmatrix} - & - & - & - \\ 0 & 1 & 0 & 0 \\ 0 & 0 & 0 & 1 \\ - & - & - & - \end{bmatrix} \times \begin{bmatrix} 00 \\ 01 \\ 10 \\ 11 \end{bmatrix}$$

$$= \begin{bmatrix} - \\ 01 \\ 11 \\ - \end{bmatrix}$$

$$= \begin{bmatrix} - \\ 1 \\ 3 \\ - \end{bmatrix} \text{ with respect to } \begin{bmatrix} 0 \\ 1 \\ 2 \\ 3 \end{bmatrix} \text{ and } \begin{bmatrix} 1 \\ 3 \\ - \end{bmatrix} \text{ with respect to } \begin{bmatrix} 1 \\ 2 \\ 3 \end{bmatrix}$$

and the final output vector being given by $\mathbf{R} \times w$, where

$$\mathbf{R} \times w = \begin{bmatrix} - & - & - & - \\ 0 & 1 & 0 & 0 \\ 0 & 0 & 0 & 1 \\ - & - & - & - \end{bmatrix} \times \begin{bmatrix} - \\ 0 \\ - \\ 1 \end{bmatrix}$$

$$= \begin{bmatrix} - \\ 0 \\ 1 \\ - \end{bmatrix} \text{ which reduces to } \begin{bmatrix} 0 \\ 1 \\ - \end{bmatrix} \text{ with respect to } \begin{bmatrix} 1 \\ 2 \\ 3 \end{bmatrix}$$

In this particular example, since the state 0 is undefined, the first row and column of each transition matrix could be deleted in which case the working would be as follows:

$$\mathbf{T}^0 = \begin{bmatrix} 1 & 0 & 0 \\ 0 & 0 & 1 \\ - & - & - \end{bmatrix}, \quad \mathbf{T}^1 = \begin{bmatrix} 0 & 0 & 1 \\ 1 & 0 & 0 \\ 0 & 1 & 0 \end{bmatrix}$$

$$q = \begin{bmatrix} 1 \\ 2 \\ 3 \end{bmatrix} = \begin{bmatrix} 01 \\ 10 \\ 11 \end{bmatrix} \text{ and } w = \begin{bmatrix} 0 \\ - \\ 1 \end{bmatrix}$$

$$\mathbf{T}^1 \times \mathbf{T}^1 \times \mathbf{T}^1 \times \mathbf{T}^0 \; (= \mathbf{R}) = \begin{bmatrix} 1 & 0 & 0 \\ 0 & 0 & 1 \\ - & - & - \end{bmatrix}$$

and hence

$$\mathbf{R} \times q = \begin{bmatrix} 1 & 0 & 0 \\ 0 & 0 & 1 \\ - & - & - \end{bmatrix} \times \begin{bmatrix} 01 \\ 10 \\ 11 \end{bmatrix} = \begin{bmatrix} 01 \\ 11 \\ - \end{bmatrix} = \begin{bmatrix} 1 \\ 3 \\ - \end{bmatrix}$$

and

$$\mathbf{R} \times w = \begin{bmatrix} 1 & 0 & 0 \\ 0 & 0 & 1 \\ - & - & - \end{bmatrix} \times \begin{bmatrix} 0 \\ - \\ 1 \end{bmatrix} = \begin{bmatrix} 0 \\ 1 \\ - \end{bmatrix}$$

There is no difficulty is obtaining the final state vector of a Mealy model by the above technique; the trouble arises in the determination of the final output vectors. The method of circumventing this is described by way of an example.

Consider the Mealy model having the state diagram of Figure 14.9 and the corresponding transition table of Table 14.10. With the input sequence of the previous example, determine the final state vector and the final output vector corresponding to an initial state vector of $\{00, 01, 10, 11\}$. In this example,

$$q = \begin{bmatrix} 00 \\ 01 \\ 10 \\ 11 \end{bmatrix}, \quad \mathbf{T}^0 = \begin{bmatrix} 1 & 0 & 0 & 0 \\ 0 & 1 & 0 & 0 \\ 0 & 0 & 1 & 0 \\ 0 & 0 & 0 & 1 \end{bmatrix}$$

$$\mathbf{T}^1 = \begin{bmatrix} 0 & 1 & 0 & 0 \\ 0 & 0 & 0 & 1 \\ 1 & 0 & 0 & 0 \\ 0 & 0 & 0 & 1 \end{bmatrix}, \quad w^0 = \begin{bmatrix} 0 \\ 0 \\ 0 \\ 1 \end{bmatrix}, \quad w^1 = \begin{bmatrix} 0 \\ 1 \\ 1 \\ 0 \end{bmatrix}$$

As before the final state vector is given by $\mathbf{T}^1 \times \mathbf{T}^1 \times \mathbf{T}^1 \times \mathbf{T}^0 \times q$ which reduces to $\{11, 11, 11, 11\}$. To find the final output vector, the *penultimate* state vector is first obtained by evaluating $\mathbf{T}^1 \times \mathbf{T}^1 \times \mathbf{T}^0 \times q$, in this case $\{11, 11, 01, 11\}$. The final output vector is then found from *either* the state diagram *or* the transition table. Note that a simultaneous check on the final state vector can be made at this point. Figure 14.9, or Table 14.10, in conjunction with the penultimate state vector yield $\{0, 0, 1, 0\}$ as the final output vector.

Sequences of States for Mealy and Moore Models

Whilst the above techniques yield the final state and output vectors, it is usually quicker to use the state diagrams, particularly when *all* the intermediate states or outputs are required.

In the Moore model example above, the sequence of state vectors can be seen to be $\{1, 2, 3\}$, $\{1, 3, -\}$, $\{3, 2, -\}$, $\{2, 1, -\}$, $\{1, 3, -\}$ corresponding to the input sequence 0, 1, 1, 1. Note that, for the states, once a known state becomes undefined, owing to a particular input, it can never return to a known state again. For the outputs, however, an undefined output *may* be followed by a known output since the outputs are dependent upon the states. In the example under consideration the sequence of output vectors is $\{0, -, 1\}$, $\{0, 1, -\}$, $\{1, -, -\}$, $\{-, 0, -\}$, $\{0, 1, -\}$.

If all possible states occur in the state diagram then the normal basis is used for the purpose of logical manipulation. However, as was seen above, one or more *may* be missing and in these circumstances it is usual to ignore the missing states when forming the transition matrix. This was done in a previous example. The manipulation is then effected with a constrained basis. This will frequently enable considerable simplification to be made to the logic circuits.

Constraints in Design

Example. Design the logic circuitry for the Mealy model having the state diagram of Figure 14.15. This is called a counter modulo 5 since there is an

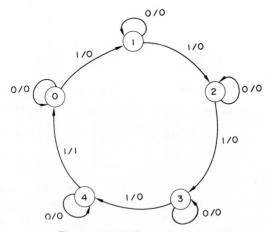

Figure 14.15 Counter modulo five

output of 1 for every fifth input of 1. Moreover, it cycles through five distinct states—in this case 0, 1, 2, 3, and 4. The use of states 0 to 4 in decimal means that three binary state variables are all that are necessary to implement this device. The decimal state table for this counter is given by Table 14.14 and it

Table 14.14

Present state	0 1 2 3 4 5 6 7	0 1 2 3 4 5 6 7
Input	0 0 0 0 0 0 0 0	1 1 1 1 1 1 1 1
Next state	0 1 2 3 4 – – –	1 2 3 4 0 – – –
Output	0 0 0 0 0 – – –	0 0 0 0 1 – – –

can be seen that six columns out of the sixteen are redundant. The equivalent binary state table of Table 14.15 is used to determine the constraint. In this

Table 14.15

Present state	z	0 1 0 1	0 1 0 1	0 1 0 1	0 1 0 1
	y	0 0 1 1	0 0 1 1	0 0 1 1	0 0 1 1
	x	0 0 0 0	1 1 1 1	0 0 0 0	1 1 1 1
Input	p	0 0 0 0	0 0 0 0	1 1 1 1	1 1 1 1
Next state	z'	0 1 0 1	0 – – –	1 0 1 0	0 – – –
	y'	0 0 1 1	0 – – –	0 1 1 0	0 – – –
	x'	0 0 0 0	1 – – –	0 0 0 1	0 – – –
Output	w	0 0 0 0	0 – – –	0 0 0 0	1 – – –

case the constraint has a designation number of

$$1 \ 1 \ 1 \ 1 \ 1 \ 0 \ 0 \ 0 \ 1 \ 1 \ 1 \ 1 \ 1 \ 0 \ 0 \ 0$$

with respect to the standard basis $b[z, y, x, p]$ and so

$$C : x(y + z) = I$$

Table 14.16

Present state	z	0 1 0 1 0 0 1 0 1 0
	y	0 0 1 1 0 0 0 1 1 0
	x	0 0 0 0 1 0 0 0 0 1
Input	p	0 0 0 0 0 1 1 1 1 1
Next state	z'	0 1 0 1 0 1 0 1 0 0
	y'	0 0 1 1 0 0 1 1 0 0
	x'	0 0 0 0 1 0 0 0 1 0
Output	w	0 0 0 0 0 0 0 0 0 1

However, a knowledge of the functional form of the constraint is incidental to the problem unless this function or its negation occurs as a part of the required functions x', y', z', and w. Using the constrained basis, as shown in Table 14.16, or by means of a Karnaugh map, in which the *don't care* conditions correspond to the missing columns of Table 14.16, the next states and the output simplify to

$$x' = pyz + \bar{p}x$$
$$y' = p\bar{y}z + \bar{p}y + y\bar{z}$$
$$z' = p\bar{x}\bar{z} + \bar{p}z$$

and

$$w = px$$

Note that, if a state diagram is generated from the above functions, and all possible states are allowed to occur, then Figure 14.16 results. From this it can

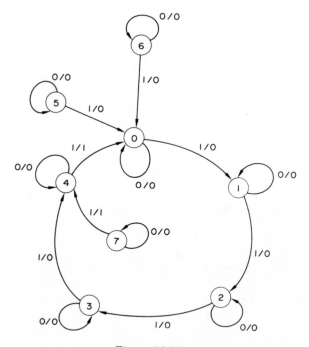

Figure 14.16

be seen that if either of the undefined states 5, 6 or 7 arise then, after a single input of 1, the counter cycle of 0, 1, 2, 3, 4 repeated must be entered. The logic circuit diagram of Figure 14.17 completes the problem.

The unconstrained basis $b[z, y, x, p]$ can be used to *force* all the undefined

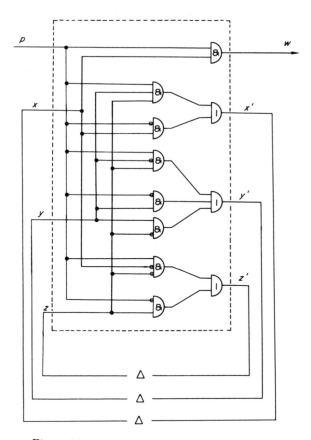

Figure 14.17 Circuit for modulo five counter

states to move to state 0. This is achieved by the state diagram of Figure 14.18 which has Table 14.17 as its associated binary state table. The next state and output functions being, in this case,

$$x' = \bar{p}x\bar{y}\bar{z} + p\bar{x}yz$$
$$y' = \bar{p}\bar{x}y + p\bar{x}\bar{y}z + p\bar{x}y\bar{z}$$
$$z' = \bar{p}\bar{x}z + p\bar{x}\bar{z}$$

and

$$w = px\bar{y}\bar{z}$$

Example 3, in the introduction to Chapter 12, asked if an existing unit (Figure 12.3) could be employed to implement the x', y', and z' of Figure 14.17.

The result of employing three of these basic units with the addition of only two negaters is shown in Figure 14.19.

Table 14.17

z	0	1	0	1	0	1	0	1	0	1	0	1	0	1	0	1
y	0	0	1	1	0	0	1	1	0	0	1	1	0	0	1	1
x	0	0	0	0	1	1	1	1	0	0	0	0	1	1	1	1
p	0	0	0	0	0	0	0	0	1	1	1	1	1	1	1	1
z'	0	1	0	1	0	0	0	0	1	0	1	0	0	0	0	0
y'	0	0	1	1	0	0	0	0	0	1	1	0	0	0	0	0
x'	0	0	0	0	1	0	0	0	0	0	0	1	0	0	0	0
w	0	0	0	0	0	0	0	0	0	0	0	0	1	0	0	0

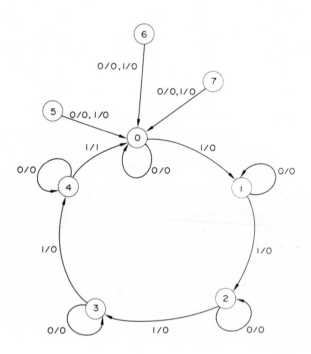

Figure 14.18

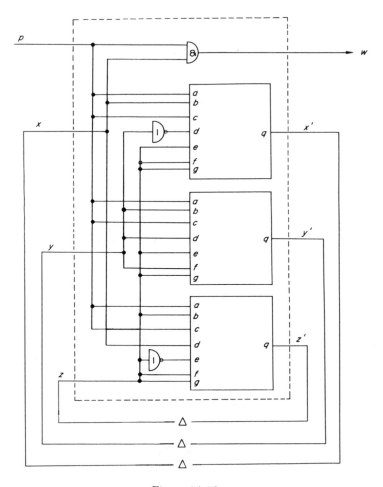

Figure 14.19

Suggestions for Further Reading

Booth, T. L., 1967, *Sequential Machines and Automata Theory*, Wiley, New York.

Miller, R. E., 1965, *Switching Theory*, Vol. 2, Wiley, New York.

Lewin, D., 1968, *Logic Design of Switching Circuits*, Nelson, London.

Torng, H. C., 1967, *Introduction to the Logic Design of Switching Circuits*, Addison-Wesley, Reading, Mass.

Nelson, R. J., 1968, *Introduction to Automata*, Wiley, New York.

Exercises

1. Using Figure 14.4, in conjunction with the table

Present state	0 1 2 3	0 1 2 3
Input	0 0 0 0	1 1 1 1
Next state	1 3 0 1	3 2 0 1
Output	0 1 0 0	0 1 0 0

    produce
      (a) the transition table
      (b) the transition matrices
      (c) the state diagram

2. The logic expressions for a simple Mealy model, having an input p and an output w, are

$$x' = \overline{\overline{x + y} + p}$$
$$y' = \overline{x + \bar{y} + p}$$
$$w = \overline{x + \bar{p}}$$

    Sketch the state diagram and produce the output sequence corresponding to an input sequence of 0 0 0 1 0 1 (reading from the left) and an initial state of 1.

3. A standard basis $b[z, y, x]$ is subject to the constraint

$$C : x\bar{y} = I$$

    Simplify, with respect to the constrained basis,
      (a) $xz + xy$
      (b) $x\bar{z} + (x\bar{y} \to z)$

4. Show that $\overline{y \to x}$ and $\overline{x \to y}$ are logically *dependent*, find the dependence relationship and determine *one* constraint which will make them independent with respect to the constrained basis.

5. A counter modulo 3 has the states 1, 3, 5 repeating, as its only permitted states, the output pulse coinciding with state 5. Produce a set of logic equations which will fit the Mealy model representation.
    Produce also a set of logic equations for the corresponding model in which any of the states 0 to 7 inclusive can occur with the proviso that if any of the states 0, 2, 4, 6, or 7 do occur then the next state will be 3 irrespective of the value of the input.

6. Given the circuit, of the type shown in Figure 14.6(a), together with the relationships

$$y' = \bar{x} + \bar{p}$$
$$x' = \overline{x + y}$$

and

$$w = \overline{x + p + \bar{y}}$$

produce
    (a) a decimal state table
    (b) the transition matrices corresponding to (a) above
    (c) the state diagram for the original states x and y

7. Using the table of Question 1 as data for a Moore model, produce the final state vector corresponding to the initial state vector $\{0, 1, 2, 3\}$ for the input sequence 0, 1, 0, 1, 0 (reading from the left).

Chapter 15
Asynchronous Applications of Delays

Asynchronous Sequential Circuits

All actual logic devices have some form of delay associated with them and this parasitic delay will be dealt with, by means of synchronization, in Chapter 16. For the present, the delays introduced into the circuitry will be more substantial and will be assumed to yield a clear-cut finite delay of known magnitude. However, even though asynchronous circuits are to be discussed, time must of necessity be taken into consideration whenever delays are involved and, for the purposes of this chapter, all delays will be measured in terms of multiples of a *unit delay* (Δ). Figure 15.1 shows the operation of delays of 1 and 3 upon an

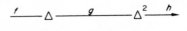

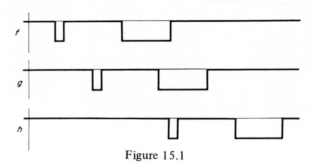

Figure 15.1

input sequence f. The outputs g and h can be referred to as Δf and $\Delta^3 f$ respectively and $h = \Delta^2 g$. Other properties of the operation Δ are those that occur *under steady state conditions,*

$$\Delta \bar{x} = \overline{\Delta x}$$

$$\Delta(x + y) = \Delta x + \Delta y$$

and

$$\Delta(x . y) = \Delta x . \Delta y$$

The second of these is illustrated in Figure 15.2.

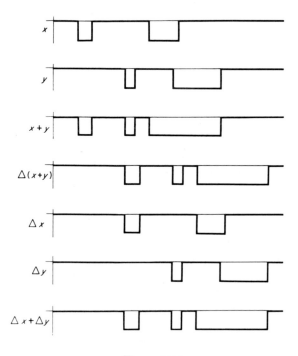

Figure 15.2

Sequence Generators

Delays used in association with feedback loops give rise to interesting sequence generators which are exceedingly valuable for control purposes. Two main types will now be considered. They both involve a steady input signal to switch them on and they both make use of an AND gate with negated feedback inputs.

The simplest type is one in which a delay n generates n 0s followed by n 1s

and Figure 15.3 illustrates three of these. In each case there is no initial run-in
before cycling starts. The sequences shown in Figure 15.3 can be represented by

01010101010101 . . . ,   0011001100110011 . . . ,   000111000111000111 . . .

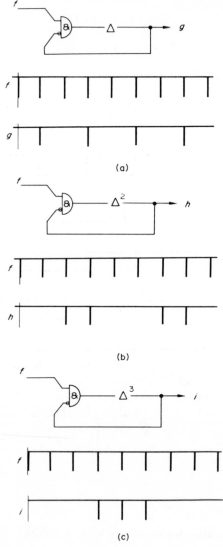

Figure 15.3

corresponding to the delays 1, 2, and 3 respectively. Note that the 1s and 0s
in the above descriptions are *not* designation numbers but are time sequences

with $t = 0$, the time at which the control signal starts, corresponding to the left-hand bit. Generators of this type can be combined in series as in Figure 15.4

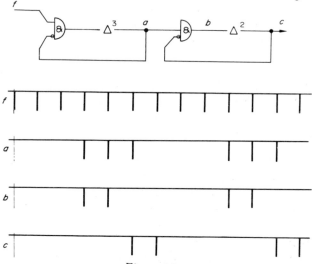

Figure 15.4

or in parallel as in Figure 15.5 to yield more complex patterns. In Figure 15.5, a and b generate respectively their steady state cycles 001111110111 and 000100000011 without an initial run-in.

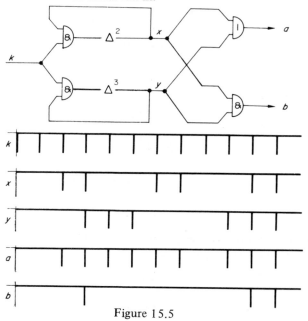

Figure 15.5

The steady state cycle of a parallel circuit of this nature is the lowest common multiple of the individual cycles, e.g. the lowest common multiple of 4 and 6 is 12 so that the steady state cycle of a circuit formed by two generators in parallel, one having a delay of 2 and the other a delay of 3, is 12.

In Figure 15.4 the generators are in series and special cases arise which need care. In general, when a delay of n is followed by a delay of m there is an initial run-in of m 0s followed by a cycle of 1s and 0s. If m is an odd multiple of n (say $m = (2r + 1)n$) then the run-in consists of m 0s and the cycle degenerates into n 0s followed by n 1s. For example for $n = 2$ and $m = 6$, the run-in consists of 000000 and the steady state cycle which then follows is 00 11 so that the complete sequence starts

$$000000001\,1001\,1001\,10011001100\ldots\ldots$$

If m is an even multiple of n (say $m = 2rn$) then the run-in can be included in the cycle which then consists of m 0s followed by r sequences of n 0s alternating with n 1s. For example for $n = 1$ and $m = 6$, the run-in, which is identical with half the cycle, is 000000. This is followed by 01 01 01, i.e. the complete cycle is 000000010101.

Using the Δ notation, Figure 15.4 can be described as follows:

$$a = \Delta^3 \bar{a}$$
$$b = a \,.\, \bar{c}$$

and

$$c = \Delta^2 b$$

therefore

$$c = \Delta^2 (a \,.\, \bar{c})$$

or

$$c = \Delta^2 a \,.\, \Delta^2 \bar{c}$$

Similarly for Figure 15.5

$$x = \Delta^2 \bar{x}$$

and

$$y = \Delta^3 \bar{y}$$

hence

$$a = \Delta^2 \bar{x} + \Delta^3 \bar{y}$$

and

$$b = \Delta^2 \bar{x} \,.\, \Delta^3 \bar{y}$$

The steady state sequences can be produced without drawing the timing diagrams as follows. From Figure 15.4

$$a = \Delta^3 \bar{a}$$

or

$$\bar{a} = \Delta^3 a$$

and
$$c = \Delta^2 \bar{c}. \Delta^2 a$$

as was shown above. Since
$$\Delta \bar{a} = \Delta^4 a$$

and
$$\Delta^2 c = \Delta^4 \bar{c}. \Delta^4 a$$
$$\Delta^2 c = \Delta^4 \bar{c}. \Delta \bar{a}$$

or
$$\Delta c = \Delta^3 \bar{c}. \bar{a}$$

Taking the logical conjunction, of each side, with a results in
$$\Delta c . a = 0$$

Therefore for $a = 0\,0\,0\,1\,1\,1$ repeated
$$\Delta c = \text{X X X 0 0 0}$$

where X indicates either a 1 or a 0, to be determined, and so
$$c = \text{X X 0 0 0 X}$$

Further
$$\Delta^2 c = 0\,\text{X X X}\,0\,0$$
$$\Delta^2 \bar{c} = 1\,\text{X X X}\,1\,1$$

and
$$\Delta^2 a = 1\,1\,0\,0\,0\,1$$

Therefore
$$\Delta^2 \bar{c}. \Delta^2 a = 1\,\text{X}\,0\,0\,0\,1$$

Thus, there are two alternatives for c, namely $1\,0\,0\,0\,0\,1$ and $1\,1\,0\,0\,0\,1$. Both of these are now checked to determine the correct sequence.

(1)
$$a = 0\,0\,0\,1\,1\,1$$
$$c = 1\,1\,0\,0\,0\,1$$
$$\bar{c} = 0\,0\,1\,1\,1\,0$$
$$\Delta^2 \bar{c} = 1\,0\,0\,0\,1\,1$$
$$\Delta^2 a = 1\,1\,0\,0\,0\,1$$
$$\Delta^2 \bar{c}. \Delta^2 a = 1\,0\,0\,0\,0\,1 \neq 1\,1\,0\,0\,0\,1$$

(2)
$$a = 0\,0\,0\,1\,1\,1$$
$$c = 1\,0\,0\,0\,0\,1$$
$$\bar{c} = 0\,1\,1\,1\,1\,0$$
$$\Delta^2 \bar{c} = 1\,0\,0\,1\,1\,1$$
$$\Delta^2 a = 1\,1\,0\,0\,0\,1$$
$$\Delta^2 \bar{c}. \Delta^2 a = 1\,0\,0\,0\,0\,1 = 1\,0\,0\,0\,0\,1$$

Hence the repeated sequence $1\,0\,0\,0\,0\,1$ for c is the correct one.

Any repetitive bit pattern can be obtained by means of these devices. If the number of bits which are repeated is even (say $2n$), then a generator which produces n 0s followed by n 1s is required whereas if the number of bits is odd (say $2n + 1$) then the generator must produce $2n + 1$ 0s followed by $2n + 1$ 1s.

Example. Produce the recurring sequence $q = 0\ 1\ 0\ 0\ 1\ 1$. This has six bits and so the basic generator is $c = 0\ 0\ 0\ 1\ 1\ 1$. Obviously, since $c = 0\ 0\ 0\ 1\ 1\ 1$,

$$\Delta c = 1\ 0\ 0\ 0\ 1\ 1$$
$$\Delta^2 c = 1\ 1\ 0\ 0\ 0\ 1$$
$$\Delta^3 c = 1\ 1\ 1\ 0\ 0\ 0$$
$$\Delta^4 c = 0\ 1\ 1\ 1\ 0\ 0$$

and

$$\Delta^5 c = 0\ 0\ 1\ 1\ 1\ 0$$

These can be combined as shown in Table 15.1 (c. $\Delta^3 c$ will not be used).

Table 15.1

Combination	Sequence					
$c + \Delta^3 c$	1	1	1	1	1	1
$c + \Delta^4 c$	0	1	1	1	1	1
$c + \Delta^5 c$	0	0	1	1	1	1
c	0	0	0	1	1	1
c . Δc	0	0	0	0	1	1
c . $\Delta^2 c$	0	0	0	0	0	1
c . $\Delta^3 c$	0	0	0	0	0	0

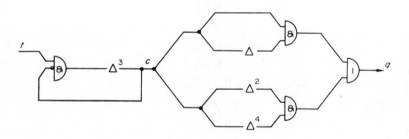

Figure 15.6

From this table of sequences, *any* six bit pattern can be formed. For example, if $q = 010011$ repeated then

$$q = [0\ 0\ 0\ 0\ 1\ 1] + [0\ 1\ 0\ 0\ 0\ 0]$$

where both sequences on the right-hand side can be constructed in a variety of ways. In particular

$$000011 = c \cdot \Delta c$$

and

$$010000 = \Delta^2(c \cdot \Delta^2 c)$$
$$= \Delta^2 c \cdot \Delta^4 c$$

Figure 15.6 shows the implementation of this particular generator from the above expressions.

Example. Produce the recurring sequence

$$p = 1 1 0 1 0 \text{ repeated}$$

In this case

$$v = \Delta^5 \bar{v}, \text{ i.e.}$$
$$v = 0 0 0 0 0 1 1 1 1 1$$

and Table 15.2 results.

Table 15.2

Combination	Sequence									
$v + \Delta^5 v$	1	1	1	1	1	1	1	1	1	1
$v + \Delta^6 v$	0	1	1	1	1	1	1	1	1	1
$v + \Delta^7 v$	0	0	1	1	1	1	1	1	1	1
$v + \Delta^8 v$	0	0	0	1	1	1	1	1	1	1
$v + \Delta^9 v$	0	0	0	0	1	1	1	1	1	1
v	0	0	0	0	0	1	1	1	1	1
$v \cdot \Delta v$	0	0	0	0	0	0	1	1	1	1
$v \cdot \Delta^2 v$	0	0	0	0	0	0	0	1	1	1
$v \cdot \Delta^3 v$	0	0	0	0	0	0	0	0	1	1
$v \cdot \Delta^4 v$	0	0	0	0	0	0	0	0	0	1
$v \cdot \Delta^5 v$	0	0	0	0	0	0	0	0	0	0

It is, of course, now necessary to generate

$$p = 1 1 0 1 0 1 1 0 1 0 \text{ repeated}$$

$$p = [1 1 0 0 0 0 0 0 0 0] + [0 0 0 1 0 0 0 0 0 0] + [0 0 0 0 0 1 1 0 0 0]$$
$$+ [0 0 0 0 0 0 0 0 1 0]$$

Hence,

$$p = \Delta^2(v \cdot \Delta^3 v) + \Delta^4(v \cdot \Delta^4 v) + \Delta^6(v \cdot \Delta^3 v) + \Delta^9(v \cdot \Delta^4 v)$$
$$= (\Delta^2 v \cdot \Delta^5 v) + (\Delta^4 v \cdot \Delta^8 v) + (\Delta^6 v \cdot \Delta^9 v) + (\Delta^9 v \cdot \Delta^{13} v)$$

However, since $v = \Delta^{10}v$ in this case

$$p = (\Delta^2v \cdot \Delta^5v) + (\Delta^4v \cdot \Delta^8v) + (\Delta^6v \cdot \Delta^9v) + (\Delta^9v \cdot \Delta^3v)$$

Alternatively, $p = (\Delta^2v \not\equiv v) + (\Delta^3v \not\equiv \Delta^4v)$ which requires only three additional delays to implement the circuit. It is this version which is shown in Figure 15.7.

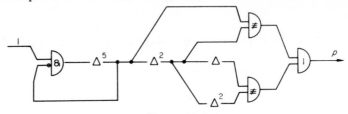

Figure 15.7

The second type of sequence generator has a number of outputs, each of which consists of a single 1 embedded in a train of 0s, and these outputs are combined to form the required pattern. Figure 15.8 shows one such sequence

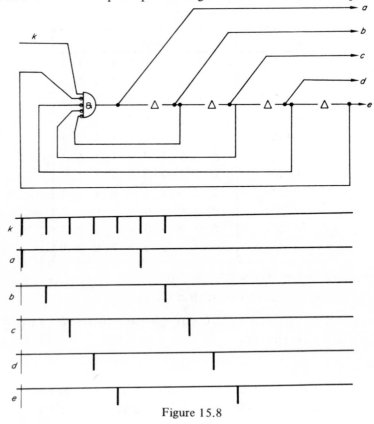

Figure 15.8

generator based upon a five-bit cycle. Note that, if the control signal k becomes 0, at any time *after* the formation of the a pulse, then the whole sequence is completed *before* the generator is switched off.

To produce the repetitive sequence $p(1\ 1\ 0\ 1\ 0)$, previously discussed, using this type of generator, it is sufficient to form the logical disjunction of its components, namely

$$p = a + b + d$$

One practical difficulty which arises with this type of generator is the large number of inputs to the AND gate when long sequences are required. One way around this is to use two such generators and form a matrix of the outputs as is shown in Figure 15.9(a) where a generator based upon a cycle length of 12 is

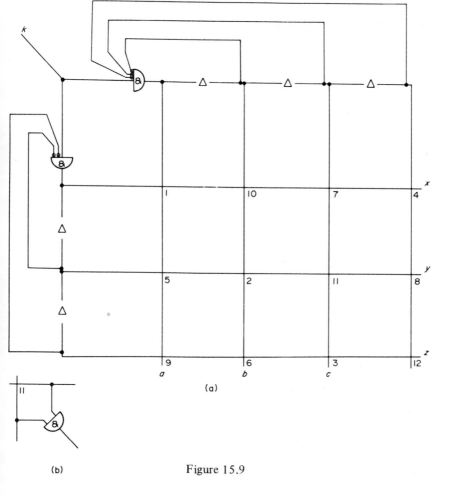

(a)

(b)   Figure 15.9

indicated. There will be a pulse appearing on the 'a' line every $4n + 1$ intervals
($n = 0, 1, 2, \ldots$), one on the 'x' line every $3n + 1$ intervals and similarly for
b, c, d, y, and z. The numbers, at the cross-over points on the diagram of
Figure 15.9(a), indicate the occurrence of a pulse if a simple AND gate is
connected as is shown in Figure 15.9(b) for junction number 11.

With such a matrix the cycle length must be the product of the cycle lengths
of its components and this may be too large, e.g. a cycle length of 11, say,
cannot be produced from such a matrix without modification. However, the
necessary modification is fairly simple. Suppose the whole system has to be
reset after the 11th pulse has been generated, i.e. the next pulse must come from
the junction labelled 1. This is achieved by taking the pulse from junction 11, by
feeding it through a unit delay, and by using the output of this delay to inhibit
the inputs to the delays which activate the 12 pulse lines. In the case of
Figure 15.9 it must inhibit the passage of pulses from c on one line and y on the
other to their following delays. The modifications necessary to Figure 15.9 are
shown in Figure 15.10.

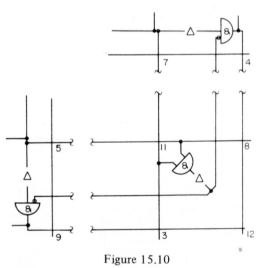

Figure 15.10

Linear Sequential Filters

A chain of NON-EQUIVALENCE units connected by delays in the form of
Figure 15.11 constitutes a *linear sequential filter*. If z is the output sequence,
corresponding to an input sequence of y, and x is the output sequence,
corresponding to an input of w, then $z \not\equiv x$ will be the output corresponding
to the input $w \not\equiv y$. Since the operations $\not\equiv$ and delay (Δ) are linear, the ratio of
the output to the input is referred to as the *digital transfer function* (d.t.f.).

In the example of Figure 15.11

$$z = y \not\equiv \Delta y \not\equiv \Delta^4 y \not\equiv \Delta^6 y$$

and writing

$$\Delta y = \Delta^1 y$$

and

$$y = \Delta^0 y$$

$$\frac{z}{y} = \Delta^0 \not\equiv \Delta^1 \not\equiv \Delta^4 \not\equiv \Delta^6$$

is the d.t.f.

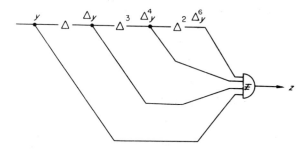

Figure 15.11

An *impulse sequence* consists of a single 1 embedded in a sequence of 0s. If the output from a linear sequential filter is obtained from an input which is an impulse sequence, then this output is called the *impulse response*.

For example, in the above case of

$$\frac{z}{y} = \Delta^0 \not\equiv \Delta^1 \not\equiv \Delta^4 \not\equiv \Delta^6$$

if

$$y = 1\,0\,0\,0\ldots$$

then

$$z = 1\,1\,0\,0\,1\,0\,1\ldots$$

In this case the seven bit sequence 'z' is the impulse response.

When z/y is of the form

$$\frac{1}{\Delta^0 \not\equiv \Delta^1 \not\equiv \Delta^4 \not\equiv \Delta^6}$$

then

$$z \not\equiv \Delta z \not\equiv \Delta^4 z \not\equiv \Delta^6 z = y$$

Since the truth of $z \not\equiv x = y$ implies that $z = x \not\equiv y$, in this case

$$z = \Delta z \not\equiv \Delta^4 z \not\equiv \Delta^6 z \not\equiv y$$

and the diagram of Figure 15.12 results. Obviously a general filter will have a transfer function in the form of a ratio, for example

$$\frac{z}{y} = \frac{\Delta^4 \not\equiv \Delta^3 \not\equiv \Delta^2 \not\equiv 1}{\Delta^3 \not\equiv 1}$$

i.e.

$$\Delta^3 z \not\equiv z = \Delta^4 y \not\equiv \Delta^3 y \not\equiv \Delta^2 y \not\equiv y$$

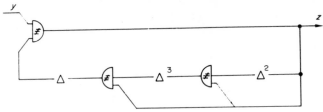

Figure 15.12

Therefore

$$z = \Delta^4 y \not\equiv \Delta^3 y \not\equiv \Delta^3 z \not\equiv \Delta^2 y \not\equiv y$$
$$= \Delta^4 y \not\equiv \Delta^3 (y \not\equiv z) \not\equiv \Delta^2 y \not\equiv y$$
$$= y \not\equiv \Delta^2 y \not\equiv \Delta^3 (y \not\equiv z) \not\equiv \Delta^4 y$$
$$= y \not\equiv \Delta^2 (y \not\equiv \Delta ((y \not\equiv z) \not\equiv \Delta y))$$

The circuit of the above is shown in Figure 15.13.

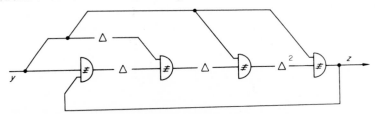

Figure 15.13

Sequence Generation Using Feedback Filters

If, for a transfer function of the form

$$\frac{z}{y} = \frac{1}{1 \not\equiv \Delta \not\equiv \ldots \not\equiv \Delta n}$$

a binary sequence exists, when y is a *null sequence*, then an output sequence has been produced *without* an input sequence.

Consider

$$\frac{z}{y} = \frac{1}{\Delta^3 \not\equiv \Delta \not\equiv 1}$$

For a null input sequence y

$$\Delta^3 z \neq \Delta z \neq z = 0$$

i.e.

$$\Delta^3 z \neq \Delta z = z \quad \text{or} \quad \Delta(\Delta^2 \neq 1)z = z$$

and Figure 15.14(a) results.

The sequence generated will depend upon the initial conditions and they in turn will depend upon the values in the delays when the devices are switched on. In the case of Figure 15.14(a) the total delay is 3 and hence there are $2^3 - 1$

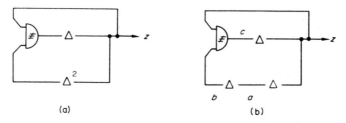

(a)   (b)

Figure 15.14

possible combinations (excluding all zeros). Suppose, for example, Figure 15.14(a) is replaced by Figure 15.14(b) and initially $a = 1$, $b = 0$, $c = 0$, so that $z = 0$. After unit delay $a = 0$, $b = 1$, $c = 1$, and $z = 0$. The resulting cycle is shown in Table 15.3 and the sequence of z will be continuously generated.

Table 15.3

a	1	0	0	1	1	1	0
b	0	1	0	0	1	1	1
c	0	1	1	1	0	1	0
z	0	0	1	1	1	0	1

Bistable Devices

The $R-S$ bistable is the only one which can be constructed without delay units; the diagram for this is the same as Figure 14.8 with the delays reduced to zero.

The main types of bistable are the T, $R-S$, $J-K$, and $R-S-T$ types. The truth tables for these are given in Tables 15.4 to 15.7 inclusive. In all cases a' is the output *before* the arrival of the input t, r, s, j, or k, as the case may be, and a is the output afterwards.

Thus for a T bistable

$$a = t \neq a'$$

and

$$\bar{a} = t \neq a'$$

Similarly for an $R-S$ bistable

$$a = (\bar{r} . a' + s)$$

and

$$\bar{a} = (\bar{s} . \bar{a}' + r)$$

together with the constraint

$$r . s = 0$$

<table>
<tr><td colspan="3" align="center">Table 15.4</td></tr>
<tr><td>t</td><td>a'</td><td>a</td></tr>
<tr><td>0</td><td>0</td><td>0</td></tr>
<tr><td>0</td><td>1</td><td>1</td></tr>
<tr><td>1</td><td>0</td><td>1</td></tr>
<tr><td>1</td><td>1</td><td>0</td></tr>
</table>

Table 15.4

t	a'	a
0	0	0
0	1	1
1	0	1
1	1	0

Table 15.5

r	s	a'	a
0	0	0	0
0	0	1	1
0	1	0	1
0	1	1	1
1	0	0	0
1	0	1	0

Table 15.6

j	k	a'	a
0	0	0	0
0	0	1	1
0	1	0	1
0	1	1	1
1	0	0	0
1	0	1	0
1	1	0	1
1	1	1	0

Table 15.7

r	s	t	a'	a
0	0	0	0	0
0	0	1	0	1
0	1	0	0	0
1	0	0	0	1
0	0	0	1	1
0	0	1	1	0
0	1	0	1	0
1	0	0	1	1

The constraint simply indicates that r and s cannot be 1 together. For a $J-K$ bistable

$$a = (\bar{k} . a' + j . \bar{a}')$$
$$\bar{a} = (k . a' + \bar{j} . \bar{a}')$$

and for an $R-S-T$ bistable

$$a = (\bar{s} \,.\, \bar{t} \,.\, a' + r \,.\, t \,.\, \bar{a}')$$

and

$$\bar{a} = (t \,.\, a' + \bar{r} \,.\, \bar{t} \,.\, \bar{a}' + s)$$

together with the constraint

$$r \,.\, s + s \,.\, t + t \,.\, r = 0$$

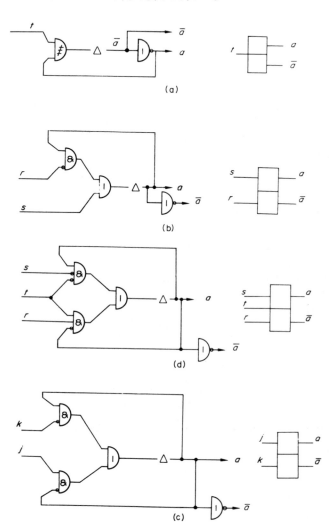

(a)

(b)

(d)

(c)

Figure 15.15

The implementation of these and their B.S. symbols are shown in Figure 15.15(a) to 15.15(d). Since the applications of these devices will be of most importance in synchronous working, a full consideration of them will be left to the next chapter.

Exercises

1. Obtain the steady state cycles for two generators in series of the type shown in Figure 15.4
   (a) when a delay of 7 is followed by a delay of 3
   (b) when a delay of 3 is followed by a delay of 7
   (c) when a delay of 3 is followed by a delay of 6
   (d) when a delay of 3 is followed by a delay of 9

2. Obtain the steady state cycles for two generators in parallel of the type shown in Figure 15.5
   (a) for delays of 3 and 5
   (b) for delays of 3 and 6

3. Design sequence generators to produce cycles of
   (a) 0 0 0 0 0 0 1 1
   (b) 0 0 0 0 0 0 0 0 0 1 0 1 0 1 0 1
   (c) 1 0 0 1 1 0
   (d) 1 0 1 1 0
   In all cases indicate whether or not the run-in can be taken to be part of the cycle.

4. Draw the circuit having the digital transfer function

$$\frac{\Delta^5 \not\equiv \Delta^3 \not\equiv \Delta^2}{\Delta^4 \not\equiv \Delta^2 \not\equiv \Delta \not\equiv 1}$$

5. Determine the impulse response of the circuit of Question 4.

Chapter 16
Parallel and Synchronous Logic

Introduction

Parallel hybrid computers are analog computers which have a certain amount of logic associated with them so that the programmer can control both individual components and the mode of the computer by digital logic, and this chapter is concerned with the application of logic to parallel hybrid computers. Nevertheless, no knowledge of analog or hybrid computers will be necessary as the emphasis will be upon the logic units and the timing considerations involved. In other words, the practical points arising from using actual units rather than theoretical devices will now be considered. The main consideration is that of synchronous or asynchronous working. All the gates so far considered will, in practice, give rise to *propagation delays* of several *nanoseconds* (1 ns = 10^{-9} s). Whilst this may not cause any difficulty in some simple logic circuits, it can mean that in others when logical levels should change together, to produce the desired result, they may not coincide in practice. In general it will be assumed that, for units of the same specification, the propagation delay is the same.

Logic Units

The timing diagram of Figure 16.1 shows the delays involved in one form of logic comparator in which the output s and its negation $\bar{s}$ should be compared with the ideal $a \equiv b$ and its negation $a \not\equiv b$. The overlap between the ideal state and reality can be seen. In this diagram, it has been assumed that the negation of the inputs does not introduce any delays whilst that of the output does.

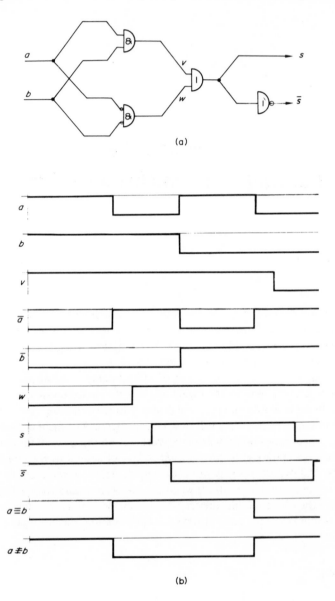

(a)

(b)

Figure 16.1

In addition to the comparator circuit described above, there is the *selector circuit* which operates under the control of a logic signal c such that the output z for two inputs a, b is given by

$$z = \begin{cases} a \text{ if } c = 1 \\ b \text{ if } c = 0 \end{cases}$$

or

$$z = ac + b\bar{c}$$

and the *complementor circuit* defined by

$$z = ac + \bar{a}\bar{c}$$

which produces an output z that is the same as the input a if $c = 1$, otherwise it gives the negation or complement of the input.

The remaining unit which does involve propagation delays but *no others* is the $R-S$ bistable, discussed in a previous chapter, in which the output changes from 0 to 1 if the *set* logic level s is 1 and from 1 to 0 if the *reset* logic level r is 1—the condition for the simultaneous occurrence of s and r both equal to 1 is indeterminate. An indeterminancy such as this is referred to as a *race condition*, since changes in voltage are racing round the circuit, and the resulting state will depend upon a number of interconnected factors which are exceedingly difficult to disentangle. A situation such as this is said to constitute a *hazard* and, obviously, hazards must be eliminated, as far as is possible, otherwise the logic devices in which they occur will yield indeterminate results.

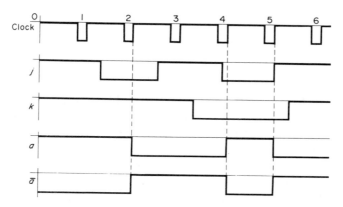

Figure 16.2

In practice, the elimination of these hazards is effected by the use of delay elements and *synchronization of operations.* This latter entails the provision of a clock, or clocks, to control the logic. Two stages are implied in synchronous working. First a *decision,* as to which outputs are going to change, must be made and secondly this decision has to be implemented. Since the *action* will always follow the *decision* it will not matter whether a two-pulse system (a decision pulse followed by an action pulse) or a single-pulse system (a decision or leading edge followed by an action or trailing edge) is adopted. In what follows the single-pulse system will be preferred and Figure 16.2 shows the effect of asynchronous (non-clocked) inputs upon a synchronous (clocked) $J-K$ bistable. Note that the output *must* of necessity be clocked. Note also that the clock pulses are *not* shown as inputs to the $J-K$ bistable nor will they be shown on any other synchronous device. In Figure 16.2 the leading edges of the clock pulses are the decision edges and hence for pulse number 4 the j input is still zero as far as synchronous working is concerned. The *only* logic levels which have any effect will be those which are active *when the leading edge of the clock pulse is present.*

Example. Figure 16.3 illustrates a five-input $a, b, c, d,$ and r and three-output $u, w,$ and z system involving two AND gates and two OR gates, all of which are

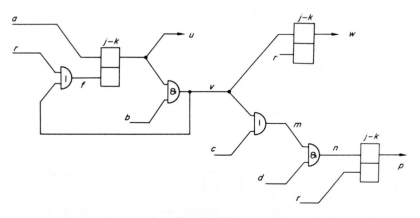

Figure 16.3

subject to propagation delays, and three $J-K$ bistables. The inputs $a, b,$ and c are asynchronous and r is a reset or clear pulse applied initially to set the outputs of all the bistables to 0. Sketch the timing diagrams for all the outputs (gates and bistables).

The timing diagrams are given in Figure 16.4 and are obtained with the reasoning shown below. The reset pulse r is 1 between the start and pulse 1 only and therefore at *decision time* 1 only. Since $f = r + v,$ f becomes 1 after a *suitable propagation delay* (SPD). Hence at decision time 1 only r and f are 1.

Therefore at action time 1 the bistables are cleared and, between pulses 1 and 2, u, w, and p are 0. c becomes 1 in this range and therefore, since $m = c + v$, m becomes 1 after an SPD. At decision time 2, and action time 2 in this case, c and m are both 1, all other signals being 0. Hence there is no change in the states of the bistables. Between pulses 2 and 3, b and c are unchanged and a and d become 1. m is also unchanged and, since $n = dm$, n will rise to 1 with d, subject only to an SPD. Thus at decision time 3, and also action time 3, a, c, d, m, and n are 1 and the others are 0. Hence the bistables' outputs u and p are set to 1, w remaining at 0. Between pulses 3 and 4, a and c

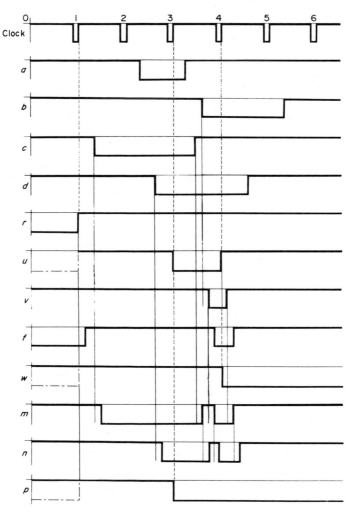

Figure 16.4

drop to 0 (but not simultaneously), *d, u,* and *p* remain 1, and *b* changes from
0 to 1. Since $v = bu$, v becomes 1 with *b* subject only to an SPD and
f ($f = r + v$) becomes 1 also, subject only to a further SPD. Because *c* becomes
0 *before* *v* becomes 1 there is a temporary fall to 0 in *m* and hence in *n* subject
only to an SPD. Therefore at decision time 4, and action time 4, *r, a,* and *c* are
0, all others being 1. Hence *p* remains 1, *w* becomes 1, and *u* changes to 0.
Between pulses 4 and 5, *r, a, b, c, u, w,* and *p* remain constant (1 or 0 as the
case may be) and *d, v, f, m,* and *n* change from 1 to 0. Hence at decision time 5,
and action time 5, *c, w,* and *p* are the only levels at 1, all others being 0. Hence
none of the bistables change. Between pulses 5 and 6, *b* drops to 0 but this
does not cause any change in any other variable and so the system settles down
in the state *w* and *p* at 1 and all others 0.

Synchronous Bistables

The asynchronous $R{-}S$ bistable can be used as the basis for all synchronous
bistables provided adequate delays can be introduced into the circuit if the
SPD is not large enough. Since most of the bistable applications which follow
will refer to the $J{-}K$ bistable rather than the $R{-}S$ bistable, the synchronous
$J{-}K$ bistable will be considered in detail. First, however, a detailed description
of the asynchronous $J{-}K$ bistable will be given as an introduction and a base
upon which to build. Figures 16.5 and 16.6 illustrate the operation of such a

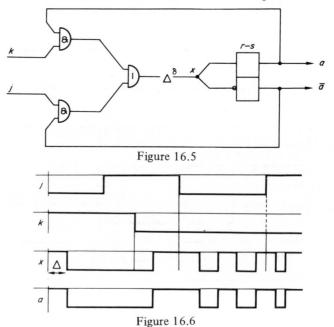

Figure 16.5

Figure 16.6

device where Δ equals the SPD of the gates, augmented by the delay δ if necessary. In this example the $R-S$ bistable has its output a at 0 initially.

In this circuit, note that if a is 1 and k is 0 then, after a delay Δ, x becomes 1 so that this constitutes a steady state independent of j, the other possible steady state occurring when a is 0 and j is 0. If a is 0 and j and k are 1 together then, after a delay Δ, x becomes 1 owing to j & $\bar{a}$ being 1. If a is 1 and j and k are 1 together then x becomes 0, because then neither of the AND gates will allow any logic signals to pass. Hence, if j and k stay at 1 for an appreciable length of time the output a will oscillate with a period of Δ.

The Synchronous $J-K$ Bistable

A circuit of a synchronous $J-K$ bistable is shown in Figure 16.7 and its timing diagram in Figure 16.8. D represents the decision pulse and A the action pulse and they are considered here as entirely separate pulse trains to avoid confusion.

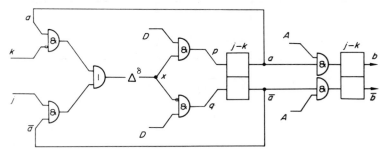

Figure 16.7

Moreover, the input signals are assumed to be synchronized to avoid cluttering up the diagrams and, as before, only asynchronous $R-S$ bistables are used in the construction. Figure 16.9 shows the effect of running the decision and action pulses into one. For all practical purposes, the trailing edge of the b outputs will coincide with the trailing edges of the action clock pulses.

Other Synchronous Bistables

A *trigger* or T bistable can be constructed from a $J-K$ bistable by simply connecting the inputs together. This is shown in Figure 16.10. Note that, if a permanent 1 is patched into *both* inputs of a $J-K$ bistable, it must act as a synchronized oscillator having a rectangular waveform and a frequency equal to one-half of that of the clock as is shown in Figure 16.11. An $R-S-T$ bistable can also be formed from the $J-K$ by means of two OR gates as shown in Figure 16.12. Note that no *two* inputs are allowed to be 1 together for an $R-S-T$ bistable.

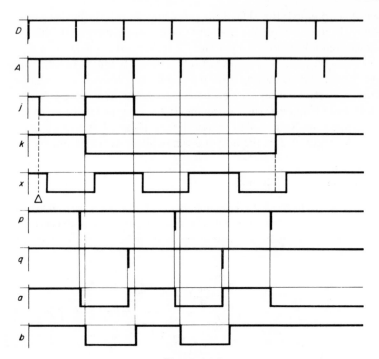

Figure 16.8

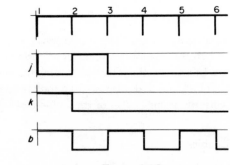

Figure 16.9

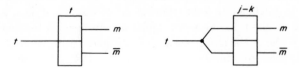

Figure 16.10

'Enabling' Controls

It is sometimes necessary to inhibit the action of a bistable from time to time and so provision is frequently made for an *enable* input, i.e. if the enable input is a 1 the bistable acts as an ordinary bistable and if the enable is a 0 its output is frozen. These enable inputs are merely additional inputs to the first pair of AND gates in Figure 16.7.

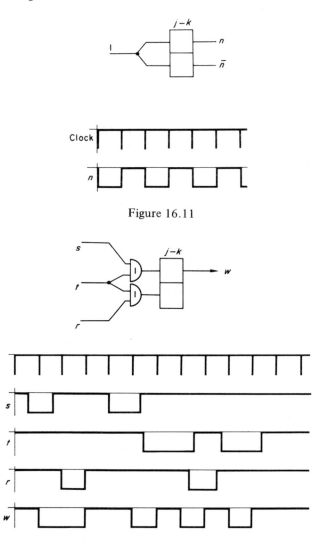

Figure 16.11

Figure 16.12

Bistable Applications

The simplest application is that of synchronizing and shaping. The circuit and timing diagrams are shown in Figure 16.13. A particularly powerful set of applications is that of the *differentiators*–leading edge, trailing edge, and

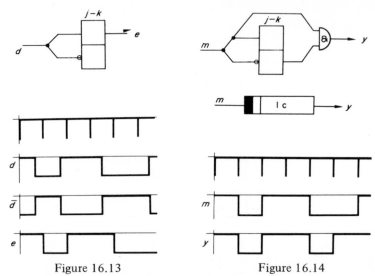

Figure 16.13   Figure 16.14

bi-polar. The formation of these devices, their B.S. symbol, where it exists, and the timing diagrams are shown in Figure 16.14, 16.15, and 16.16 respectively. The symbol for the bi-polar differentiator is non-standard as there is no B.S. symbol for this. Further, its leading edge output merges into its trailing edge,

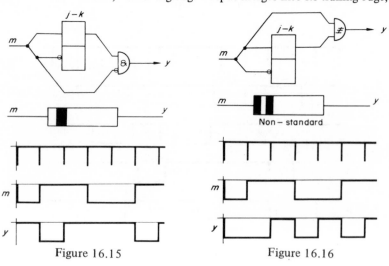

Figure 16.15   Figure 16.16

irrespective of the length, whenever the input is a single clock interval in length and the trailing edge of one differential merges into the leading edge of the next if the separation of the two input pulses is one clock interval.

A simple delay of one clock interval can also be realized by means of a single J–K bistable as is shown in Figure 16.17.

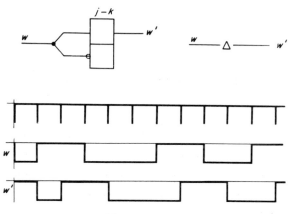

Figure 16.17

Sampling, and synchronizing, can be carried out by means of the enable input to the synchronized J–K bistable as is shown in Figure 16.18. Note that, owing to the absence of the enable pulse, input pulses a and c are ignored.

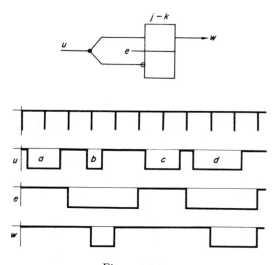

Figure 16.18

A *monostable* device requires some additional clocked delays and produces
an output of a fixed number of clock intervals for every input. The circuit and
timing diagrams are as in Figure 16.19 for a delay of 1 clock interval (1 c).

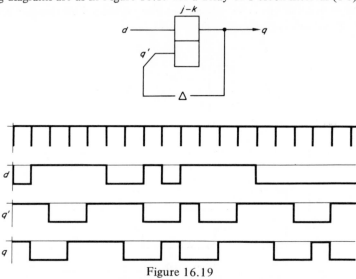

Figure 16.19

This yields an output pulse of width two clock intervals and in general a delay
of *n* clock intervals will yield an output pulse width of *n* + 1 clock intervals.
Notice in particular that when the input is held constant at 1 the output
consists of a train of pulses of the requisite width separated by unit clock

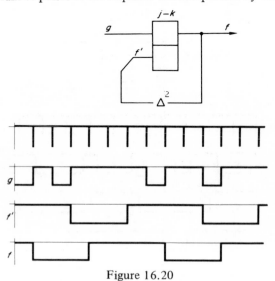

Figure 16.20

intervals for which the output is 0. This *down time* is constant at one clock interval irrespective of the output pulse width. If the input pulses change from 1 to 0 and then from 0 to 1 again whilst the output is a 1 then they are ignored as can be seen from Figure 16.20 which involves a delay of two clock pulses.

A Long-term-change Detector

The circuit of Figure 16.21 produces the effect of detecting long-term changes in logic level. Note that the enable input is applied to the AND gate *as well as to the bistable* and thus the output becomes 1 every time an input change coincides with an enable pulse.

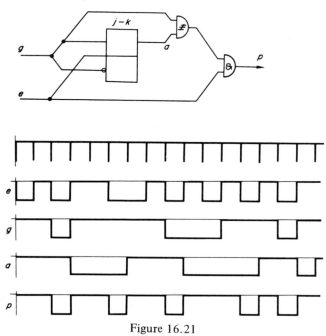

Figure 16.21

General Purpose Registers (GPR)

The symbol for these devices is shown in Figure 16.22(a) for the case where four bits of information are to be stored. A, B, C, and D identify the individual bit locations and the outputs a, b, c, and d will therefore each carry a 1 or a 0, d being the least significant. Figure 16.22(b) shows a *common clear* input which, when 1, zeros all the bits. Obviously individual clear signals can be used as is shown in Figure 16.22(c) when r_b clears B only, r_d clears D only and r clears A, C, and D. The letters identifying the bit

locations may be omitted if no ambiguity is likely to arise. Figure 16.22(d)
shows a *common set* input and Figure 16.22(e) one in which the bit pattern
1010 is established. Figure 16.23 shows a register with a common clear line and
individual set and *read-out* lines. If the register is cleared *before* the set lines
are actuated then any general bit pattern can be coded and, since the register
consists essentially of a chain of bistables, read-out can take place at any time.

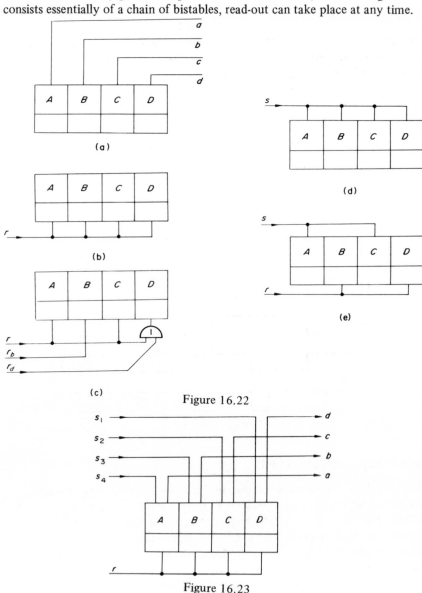

Figure 16.22

Figure 16.23

Shifting

Shifting left, i.e. moving the bit pattern towards the most significant end and feeding a 0 into the least significant end, is effected by a *step* signal, usually the enable signal, input at the r.h.s., the arrow indicating the direction of shift. This

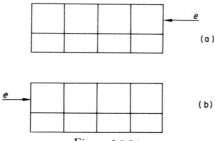

Figure 16.24

is shown in Figure 16.24(a) and the corresponding *shift right* by a step signal input at the l.h.s. as in Figure 16.24(b). Tables 16.1 and 16.2 illustrate the effect of shifting as a bit pattern a, β, γ, δ (each Greek letter can be 1 or 0).

Table 16.1

a	b	c	d	
0	α	β	γ	shift right
0	0	α	β	
0	0	0	α	
0	0	0	0	

Table 16.2

a	b	c	d	
α	β	γ	0	shift left
β	γ	0	0	
γ	0	0	0	
0	0	0	0	

Cyclic stepping action is indicated by a curved arrow on one side or the other as required. Cyclic stepping simply means that any bits shifted out of the register at one end are automatically fed back into the other end. The symbolic diagram for this is shown in Figure 16.25(a). Note that it is *not* necessary to carry the arrow right round the register to the other end. Serial

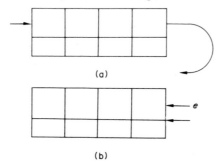

(a)

(b)

Figure 16.25

input to a register takes place when, on every step commenced, a shift left
takes place and a new bit is fed into the least significant end, instead of the 0
as in a simple shift left. These bits will normally come from other registers,
being shifted out, or from bistables or arithmetic units. The schematic
diagram is shown in Figure 16.25(b). The construction of the individual components of element B of one of these shift left registers is shown in Figure 16.26

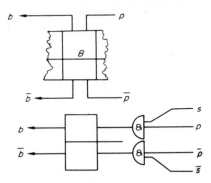

Figure 16.26

where p is the input from the previous stage. Note that Figure 16.26 will allow
shifting to the right when p is the output of A, the stage to the left of B, and to
the left when p is the output of C, the stage to the right of B. Hence the circuit
of Figure 16.26, for element B, will serve both as a shift left and a shift right
register if the input p to element B is replaced by

$$ar + cl$$

where r is a right shift control signal and l is a left shift control signal. When a
shift pulse is present $r = \bar{l}$. If the register is connected to cycle, then the output
from the most significant end is fed back to the least significant end, otherwise
there is no input.

Figure 16.27(a) shows a cyclic shift left GPR with an initialization input k,
in this case setting the pattern 0001, and an output q every subsequent fourth
step (enable) pulse. Any four-bit pattern can of course be set initially into
such a register. However, with this particular pattern it acts as a counter
modulo-4 as can be seen from the timing diagram of Figure 16.27(b) so that
the B.S. symbol of Figure 16.27(a) can be replaced, for reasons of economy,
with the symbol shown in Figure 16.27(c).

Counting

In addition to the counter modulo-4 designed above, and in fact all counters
modulo $2n$ obtained by adding $J–K$ bistables to the end of the above set of

four, counters following specific bit patterns can be constructed, such as BCD counters. A *synchronous binary up counter* generates the binary equivalents of the decimal digits 0 to 15 in succession and then repeats them. One method is

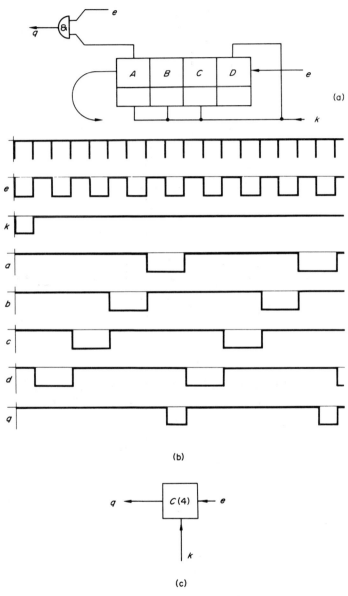

Figure 16.27

to connect the bistables of a register as T bistables when the schematic symbol is drawn as in Figure 16.28(a); $a, b, c,$ and d being weights 8, 4, 2, and 1 respectively. Two alternative forms are shown in Figure 16.28(b) and (c).

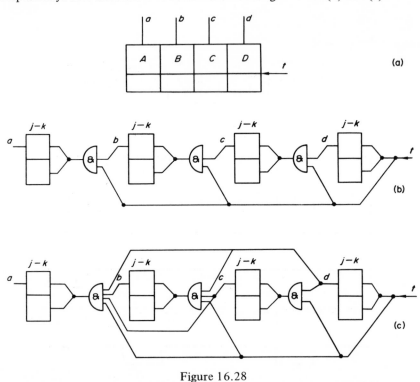

Figure 16.28

To produce a *BCD up counter* from the binary counter of Figure 16.28(a) the counter must be reset after ten input pulses and a logic 1 transmitted from the unit. This means counting 0, 1, 2, 3, 4, 5, 6, 7, 8, 9, 0 with a 1 being transmitted when the binary pattern corresponding to decimal 9 occurs, i.e. immediately a and d are 1 together the bistables must clear. Figure 16.29 shows how this is achieved. The negated inputs to the AND gate are not logically necessary but they constitute a fail-safe device.

A Johnson counter can be constructed quite simply if an additional bistable is employed since the code is a five-bit code. The sequence is

f sets d	2
d sets c	3
c sets b	4
b sets a	5
a sets $\bar{f}$ (resets f)	6

$$\bar{f} \text{ sets } \bar{d} \text{ (resets } d) \quad 7$$
$$\bar{d} \text{ sets } \bar{c} \text{ (resets } c) \quad 8$$
$$\bar{c} \text{ sets } \bar{b} \text{ (resets } b) \quad 9$$
$$\bar{b} \text{ sets } \bar{a} \text{ (resets } a) \quad 0$$
$$\bar{a} \text{ sets } f \quad 1$$

repeated.

Thus, for element B, using $J-K$ bistables, the set input j_B is equal to c and the reset input k_B is equal to $\bar{c}$. The step inputs coincide with the enable pulses as before. Figure 16.30 shows the timing.

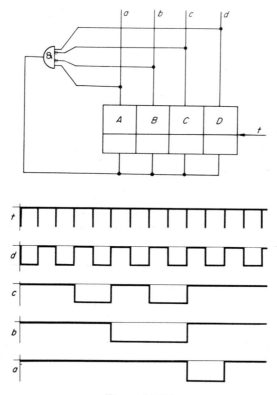

Figure 16.29

Timing of Repetitive Operations

It is usual to have custom-built counters which can be used as timing devices. Whilst these counters can be either *up counters* or *down counters* the latter are most frequently met in practice. These devices give an output pulse of 1 when the counter reaches zero. The total count is usually pre-set in some manner

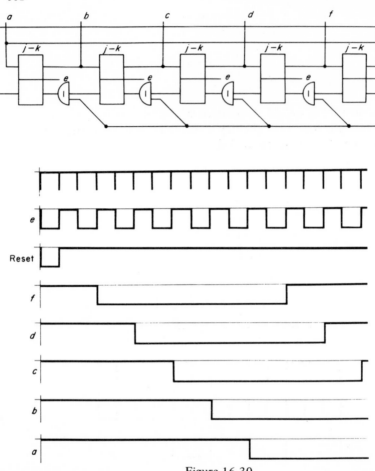

Figure 16.30

such as hand switches or push buttons. The symbol used will be as in Figure
16.31 where the n indicates that the counter runs from $n - 1$ to 0, i.e. it counts
n pulses.

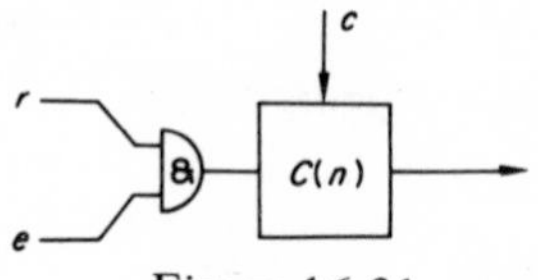

Figure 16.31

Figure 16.32 shows a hybrid *logic timer*. The inputs are the stepping pulses
synchronized with enable signals and a run logic level. The circuit lacks complete
symmetry owing to the inclusion of AND and OR gates. The OR gate ensures
that the C(4) counter is held in the *clear* condition so long as the *run* signal r

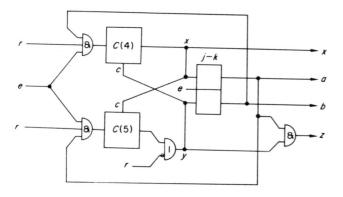

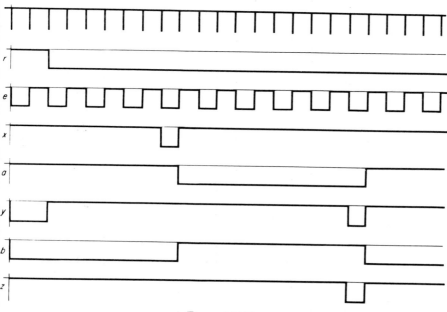

Figure 16.32

is 0 (this automatically holds $a = 0$ and $b = 1$). The AND gate produces an output z when a and y are both 1s, i.e. it signals the end of the C(5) run and the beginning of the next C(4) run. The x output signals the end of the C(4) run and the beginning of the C(5) run.

The above system is essentially a two-state system whereby an analog computer can be controlled to cycle between its *initial condition* state and its *run state*. Sometimes more sophisticated timing concepts are required and three

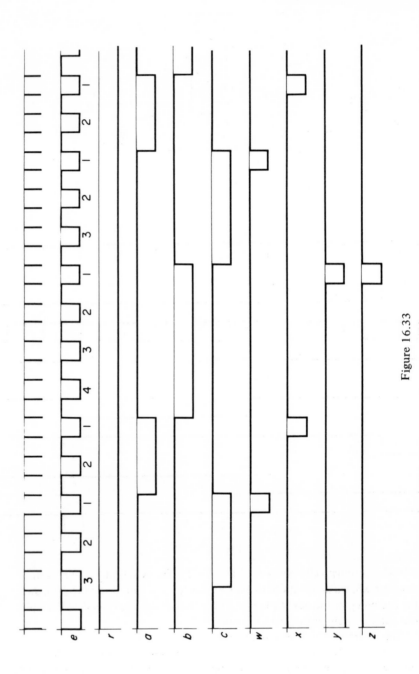

Figure 16.33

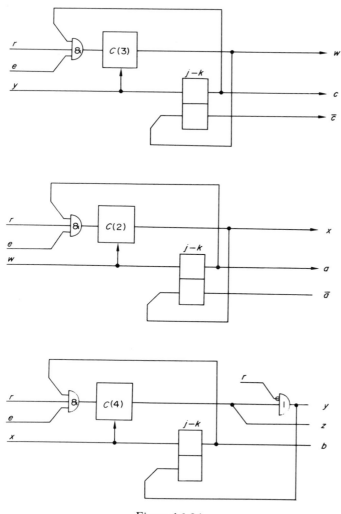

Figure 16.34

or more states may be needed. Suppose the timing diagrams of Figure 16.33 is required. The approach is based upon that of Figure 16.32, namely three bistables are used and three counters as in Figure 16.34 where r is the run logic level of 1 and e the stepping pulses, synchronized with the enable signals, a, b, and c are the required outputs of the bistables and a becomes 1 after C(3) has generated a carry, b becomes 1 after C(2) has generated a carry, and c becomes 1 after C(4) has generated a carry. w, x, and y (and hence z) initiate the start of counters C(2), C(4), and C(3) respectively—they must therefore clear these counters as well as set a, b, and c respectively. Further, when $r = 0$ both y and c must be 1 and z is the output from $C(4)$ only, a, b, and c are reset by x, y, and w respectively. To avoid a large number of crossing connections, the convention is adopted that the w, x, and y on the r.h.s. of Figure 16.34 are assumed to be fed back to the w, x, and y on the l.h.s.

Solutions

Chapter 1

1. (a) $(110\ 100\ 101)_2$, $(645)_8$
   (b) $(545)_8$, $(357)_{10}$
   (c) $(100\ 000\ 111\ 101)_2$, $(31414)_5$, $(2109)_{10}$

2. (a) $(0{\cdot}100\ 111\ 00\ldots)_2$
   (b) $(0{\cdot}7109375)_{10}$
   (c) $(0{\cdot}514007568359375)_{10}$

3. $\pi = (11{\cdot}001\ 001\ 000\ 1)_2$
   $e = (10{\cdot}101\ 101\ 111\ 0)_2$

4. $(1101{\cdot}101\ 0010\ 0010\ 0010\ldots)_2$

5. (a) $(0{\cdot}3083333\ldots)_{10}$
   (b) $(0{\cdot}4777777\ldots)_{16}$

6. $\frac{1}{2} = 0{\cdot}1$ exactly

   $\frac{1}{3} = 0{\cdot}01\ \ 01\ldots$ repeated

   $\frac{1}{4} = 0{\cdot}01$ exactly

   $\frac{1}{5} = 0{\cdot}0011\ \ 0011\ldots$ repeated

   $\frac{1}{6} = 0{\cdot}001\ \ 01\ \ 01\ldots$ repeated

   $\frac{1}{7} = 0{\cdot}001\ \ 001\ldots$ repeated

   $\frac{1}{8} = 0{\cdot}001$ exactly

$\frac{1}{9} = 0.000111\ \ 000111 \ldots$ repeated

$\frac{1}{10} = 0.0001\ \ 1001\ \ 1001 \ldots$ repeated

$\frac{1}{11} = 0.0001011101 \ldots$ repeated

7. $(0.1)_{10}$ $= 0.000\ \ 110\ \ 011\ \ 001\ \ 100\ \ 110\ \ 011$
   $(0.02)_{10}$ $= 0.000\ \ 001\ \ 010\ \ 001\ \ 111\ \ 010\ \ 111$
   $(0.005)_{10} = 0.000\ \ 000\ \ 010\ \ 100\ \ 011\ \ 110\ \ 101$

   $\Sigma\ = 0.000\ \ 111\ \ 111\ \ 111\ \ 111\ \ 111\ \ 111$
   $= (0.001)_2$

8. (a) 1 010 001   11 101
   (b)   44 065   35 575
   (c) 1 000·11111   100·01001

9. (a) 11·00   (4 significant bits)
   (b)   5·774   (4 significant bits)
   (c)   0·000 1011   (4 significant bits)

10. 11·010 (3 bits)

11. (a) (i)   1·00111
        (ii)   Answer out of range
        (iii) 1·00000 (− zero)
        (iv)   Answer out of range
        (v)   11·00111
        (vi)   1·01010
    (b) Answers as above except for (iii) which is now 0·00000 (+ zero)

Chapter 2

1. (i)   +3, −3
   (ii)   A, E, I, O, U
   (iii) Empty
   (iv)   Paris
   (v)   2

2. (a) False; −3   (b) True
   (c) False; 1   (d) False; empty
   (e) True   (f) False; −3
   (g) False; $\{2\}$ is a set not an even number (h) True
   (i) False; −2   (j) True
   (k) False; any integer   (l) False; $\bar{z}$ is not a set

3. (a) True; order of elements does not matter
   (b) False; $4 \notin \{1, 2, 3, 5\}$
   (c) True; same set of elements
   (d) False; the set $\{1, 2\}$ is not an element of the list $1, 2, \{1, 2, 3\}, \{\{1, 2\}\}$

6. $x - y = x \cap \bar{y}$
   Not commutative.
   Distributive from the right over intersection but not from the left;
   i.e. $(y \cap z) - x = (y - x) \cap (z - x)$ but $x - (y \cap z) \neq (x - y) \cap (x - z)$.
   Not associative.

8. (a) 2, 4
   (c) 3, 4, 7, 8
   (e) 1, 2, 3, 4
   (g) 1, 2, 3, 4, 5, 6, 7, 8

   (b) 2, 4, 5, 7
   (d) Empty
   (f) 1, 2, 3, 4
   (h) 3, 4

10. (a) $(w \cap x \cap y) \subset z$
    (b) $z \subset x \cup y$
    (c) $x \cap z \subset y$
    (d) $x \subset y \cap z$
    (e) $x \subset y \subset z$

12. Total number in $I = 18$

13. (a) 14%; (b) $16\frac{2}{3}\%$; (c) 60%

14. Total number of pupils $= 38$
    (a) 7; (b) 24; (c) 7

15. Number of married men $= 16$
    Number of married graduates between 1 and 10
    Number of single women non-graduates between 20 and 22

Chapter 3

4. (a) $\bar{x} \cdot (\bar{y} + \bar{z})$
   (b) $(\bar{x} + \bar{y}) \cdot (\bar{y} + \bar{z}) \cdot (\bar{z} + \bar{x})$
   (c) $x + y + z$
   (d) $x \cdot y \cdot z + x + y + z$

5. (a) x
   (b) x
   (c) $x \cdot y \cdot \bar{z}$

9. $F(x, y, z) = F(0, 0, 0) \cdot \bar{x} \cdot \bar{y} \cdot \bar{z} + F(0, 0, 1) \cdot \bar{x} \cdot \bar{y} \cdot z + F(0, 1, 0) \cdot \bar{x} \cdot y \cdot \bar{z}$
$+ F(0, 1, 1) \cdot \bar{x} \cdot y \cdot z + F(1, 0, 0) \cdot x \cdot \bar{y} \cdot \bar{z} + F(1, 0, 1) \cdot x \cdot \bar{y} \cdot z$
$+ F(1, 1, 0) \cdot x \cdot y \cdot \bar{z} + F(1, 1, 1) \cdot x \cdot y \cdot z$

Boolean function is $\bar{x} \cdot y \cdot z + x \cdot \bar{y} \cdot z + x \cdot y \cdot \bar{z} + x \cdot y \cdot z$.

10. (a) $\bar{x} \cdot \bar{y} \cdot \bar{z} + \bar{x} \cdot \bar{y} \cdot z + x \cdot \bar{y} \cdot \bar{z} + x \cdot \bar{y} \cdot z + x \cdot y \cdot \bar{z} + x \cdot y \cdot z$
(b) $w \cdot x \cdot y \cdot z$
(c) $\bar{x} \cdot y \cdot \bar{z} + \bar{x} \cdot y \cdot z + x \cdot \bar{y} \cdot z + x \cdot y \cdot z$
(d) $x \cdot y \cdot \bar{z}$

11. (a) $(x + \bar{y} + z) \cdot (x + \bar{y} + \bar{z})$
(b) All terms of the complete normal form except for $(\bar{w} + \bar{x} + \bar{y} + \bar{z})$,
i.e. 15 terms
(c) $(x + y + z) \cdot (x + y + \bar{z}) \cdot (\bar{x} + y + z) \cdot (\bar{x} + \bar{y} + z)$
(d) All terms of the complete normal form except for $(\bar{x} + \bar{y} + z)$, i.e. 7 terms

Chapter 4

1. (a) No, it is a question not a statement
(b) Yes, although its truth value is 'false' (elementary proposition g)
(c) No, there is no predicate
(d) Yes, conjunction (s & m)
(e) Yes, implication ($f \rightarrow e$)
(f) Yes, exclusive disjunction or non-equivalence ($g \not\equiv s$)
(g) Yes, if it is considered as an assertive statement and not as a command
(elementary proposition s)
(h) Yes, equivalence and conjunction ($p \equiv (m$ & $t)$)

2. (a) Today is cold and it is raining
(b) If it is raining then today is cold
(c) If yesterday was sunny then today is not cold and it is not raining
(d) Today it is neither cold nor raining if and only if yesterday was sunny
(e) It is false that today is cold and it is raining and that yesterday was
sunny
(f) Today it is either cold or it is raining, but not both, and yesterday it
was not sunny
The above answers have been written as almost direct translations. They
could be expressed in better English while still keeping the same meaning,
for example
(e) Today it is warm and dry but yesterday was not sunny

3. Taking the variables in alphabetical order, the final columns of the truth tables written from left to right instead of top to bottom are as follows:

(a) 1 0 0 1

(b) 1 0 0 0 0 0 0

(c) 1 1 0 1 1 1 0 1 0 1 0 1 1 1 0 1

(d) 0 0 0 0 0 0 0 0

4. (a) SDNF = $\bar{x}$ & $\bar{y}$ V x & y
   SCNF = (x V $\bar{y}$) & ($\bar{x}$ V y)

(b) SDNF = $\bar{x}$ & $\bar{y}$ & z
   SCNF = (x V y V $\bar{z}$) & (x V $\bar{y}$ V z) & (x V $\bar{y}$ V $\bar{z}$) & ($\bar{x}$ V y V z)
   & ($\bar{x}$ V y V $\bar{z}$) & ($\bar{x}$ V $\bar{y}$ V z) & ($\bar{x}$ V $\bar{y}$ V $\bar{z}$)

(c) SDNF = $\bar{w}$ & $\bar{x}$ & $\bar{y}$ & $\bar{z}$ V $\bar{w}$ & $\bar{x}$ & $\bar{y}$ & z V $\bar{w}$ & $\bar{x}$ & y & z
   V $\bar{w}$ & x & $\bar{y}$ & $\bar{z}$ V $\bar{w}$ & x & $\bar{y}$ & z V $\bar{w}$ & x & y & z
   V w & $\bar{x}$ & $\bar{y}$ & z V w & $\bar{x}$ & y & z V w & x & $\bar{y}$ & $\bar{z}$
   V w & x & $\bar{y}$ & z V w & x & y & z
   SCNF = (w V x V $\bar{y}$ V z) & (w V $\bar{x}$ V $\bar{y}$ V z) & ($\bar{w}$ V x V y V z)
   & ($\bar{w}$ V x V $\bar{y}$ V z) & ($\bar{w}$ V $\bar{x}$ V $\bar{y}$ V z)

(d) SDNF = 0
   SCNF = ($\bar{w}$ V $\bar{x}$ V $\bar{y}$) & ($\bar{w}$ V $\bar{x}$ V y) & ($\bar{w}$ V x V $\bar{y}$) & ($\bar{w}$ V x V y)
   & (w V $\bar{x}$ V $\bar{y}$) & (w V $\bar{x}$ V y) & (w V x V $\bar{y}$) & (w V x V y)

6. *Hint.* Show that it is not possible to produce the NOT connective, i.e. it is not possible to obtain $\bar{x}$ using & only.

8. Rich people are not students.
   If x is even ($\bar{z}$), = 2n say, then $x^2 = 4n^2$ which is even, i.e. $\bar{y}$. So that $\bar{z} \to \bar{y}$. Hence $y \to z$.

9. Let x be the proposition '*a* is a positive integer'
   y be the proposition 'a^2 is a positive integer'
   Then $x \to y$. But y does not imply x (e.g. $a^2 = 4$ is satisfied by $a = -2$; or $a^2 = 2$ for which a is not an integer). A case for which the converse does hold is the example in Question 8. Many other examples can be obtained from elementary geometry, e.g. Pythagoras' theorem.

10. (a) Valid
    (b) False
    (c) Valid

11. (a) Valid $z \to y$ premise
    $\bar{y} \to \bar{z}$ contrapositive
    $x \to \bar{y}$ premise
    $x \to \bar{z}$ syllogism
    $z \to \bar{x}$ contrapositive

$$\frac{z}{\overline{x}} \qquad \text{premise}$$
$$\overline{x} \qquad \text{detachment}$$

(b) Valid $\quad x \to y \quad$ premise

$\overline{y} \to \overline{x} \quad$ contrapositive

$\overline{x} \to z \quad$ premise $(= x \lor z)$

$\overline{y} \to z \quad$ syllogism

$\overline{z} \to y \quad$ contrapositive

$\overline{z} \qquad$ premise

$y \qquad$ detachment

(c) Valid $\quad z \to \overline{y} \quad$ premise

$y \to \overline{z} \quad$ contrapositive

$x \to y \quad$ premise

$x \to \overline{z} \quad$ syllogism

$\overline{z \to \overline{x}} \quad$ contrapositive

12. (a) $s \to k \lor q$

$k \to r$

$s \to \overline{q}$

$\overline{s \to r} \quad$ (valid)

(b) $c \to t \& p$

$\overline{r} \to \overline{t}$

$p \to r$

$\overline{c \to r} \quad$ (valid)

Chapter 5

5. (a) $x + \overline{y}$
   (b) $w . [x . (y + z) + y . z]$
   (c) $w . x$

11. $f = (w + \overline{x} + y + \overline{z}) . (\overline{w} + \overline{x} + y + z)$
    $= \overline{x} + y + (w + \overline{z}) . (\overline{w} + z)$

Chapter 6

9. (a) $([x \downarrow 0] \downarrow y) \downarrow ([y \downarrow 0] \downarrow z) \downarrow (x \downarrow [z \downarrow 0])$
   (b) $\{([p \downarrow 0] \downarrow [q \downarrow 0] \downarrow [r \downarrow 0]) \downarrow (p \downarrow q) \downarrow [r \downarrow 0]\} \downarrow 0$
   (c) Expression $= (\overline{x} + \overline{y} + \overline{z}) + (x + y + z) = 1 = 0 \downarrow 0$
   (d) $\{[p \downarrow 0] \downarrow q \downarrow [x \downarrow 0] \downarrow y\} \downarrow \{([p \downarrow 0] \downarrow q) \downarrow ([x \downarrow 0] \downarrow y)\}$

11. (a) $((((\overline{w}) . (\overline{x})) . (\overline{y})) . (\overline{z}))$
    1 2 3 4 3   4 3 2   3 2 1   2 1 0

   (b) $((((w . x) . y) + (((\overline{w}) . x) . y)) + ((w . (\overline{x})) . y))$
    1 2 3 4   3   2   3 4 5 4   3   2 1   2 3   4 3 2   1 0

   (c) $((((((w + x) + (y . (\overline{w})))) + x) + (y . w)) + (\overline{x})) + y)$
    1 2 3 4 5 6   5   6   7 6 5 4   3   4   3 2   3 2 1   0

   (d) $(((\overline{w + x})) . ((((\overline{y + z})) . ((\overline{w + z}))))))$
    1 2 3   2 1   2 3 4 5   4 3   4 5   4 3 2 1 0

Chapter 7

1. (a) 71
   (b) 114
   (c) 3
   (d) −19
   (e) −9
   (f) 271

2. (a) $((1 + 2) \times 3 + 4) \times 5 + 6$
   (b) $(2 \times (3 + 4) + 5) \times 6$
   (c) $6 - (5 - (4 - (3 - (2 - 1))))$
   (d) $1 - 2 - 3 - 4 - 5 - 6$
   (e) $6 - 5 - 4 - 3 - 2 - 1$
   (f) $6 \times (5 + 4) \times (3 + 2) + 1$

5.   Forward Polish   Reverse Polish
   (a) early V V V V xy & z N x N y N z xy V zx N & V y N V z N V
       late   V x V y V & z N x V N y N z $xyzx$ N & y N z N V V V V
   (b) early & V V xyz V V N x N y N z xy V z V x N y N V z N V &
       late   & V x V yz V N x V N y N z xyz V V x N y N z N V V &
   (c) early V & xy & yz xy & yz & V
       late   same as early operator form
   (d) early & & & & N w N & N x N & N y N z xyz
                       w N x N y N z N & N & N & x & y & z &
       late   & N w & N & N x N & N y N z & x & yz
                       w N x N y N z N & N & N xyz & & & &

6. (a) w N x N y N z N & & &
   (b) w N x N & y N & z N &
   (c) wx V y V z V $zyxw$ & & & &
   (d) $wx \rightarrow yz \rightarrow \rightarrow$

7. (a) $\overline{w} \& \overline{x} \& \overline{y} \& \overline{z}$
   (b) $\overline{w} \& \overline{x} \& \overline{y} \& \overline{z}$
   (c) $(w \lor x \lor y \lor z) \& z \& y \& x \& w$
   (d) $(w \to x) \to (y \to z)$

8. (a) $x \, N \, y \, N \, \lor$
   (b) $xy \, N \, \& \, yz \, \& \, \lor \, x \, N \, z \, N \, \& \, \lor$ (several alternatives possible)
   (c) $xy \, N \, K$

Chapter 8

1. (a) 00000000 00000000 11111111 11111111 00000000 00000000
       11111111 11111111
   (b) 0101 0101
   (c) 0011 0011 0011 0011

2. (a) 0011
   (b) 0000 0111
   (c) 1110
   (d) 0111

3. (a) SDNF $= y\overline{z} + y\overline{z}$
       SCNF $= (y + z) \& (y + \overline{z})$
   (b) SDNF $= x\overline{y}z + xy\overline{z} + xyz$
       SCNF $= (x + y + z)(x + y + \overline{z})(x + \overline{y} + z)(x + \overline{y} + \overline{z})(\overline{x} + y + z)$
   (c) SDNF $= \overline{y}\overline{z} + \overline{y}z + y\overline{z}$
       SCNF $= \overline{y} + \overline{z}$
   (d) SDNF $= \overline{y}z + y\overline{z} + yz$
       SCNF $= y + z$

4. (a) O, contradiction
   (b) I, tautology
   (c) 1111 1101, neither
   (d) I, tautology
   (e) I, tautology
   (f) 1101, neither

5. (a) Valid
   (b) Valid
   (c) Not valid
   (d) Valid
   (e) Valid

6. (a) $\bar{x}\bar{y}\bar{z} + \bar{x}\bar{y}z + \bar{x}yz + xyz$
   (b) $x\bar{y}\bar{z} + xyz$
   (c) I, all eight terms
   (d) I, all four terms

7. (a) 1000 0000, $\bar{x}\bar{y}\bar{z}$
   (b) 0111 0000, $x\bar{y}\bar{z} + \bar{x}yz + xy\bar{z}$
   (c) 0010 1010, $\bar{x}y\bar{z} + \bar{x}\bar{y}z + \bar{x}yz$

8. (a) $\bar{y}z + yz$
   (b) $uvwxyz$
   (c) $\bar{z}\bar{y}\bar{x}\bar{b}\bar{a} + \bar{z}\bar{y}\bar{x}\bar{b}a + \bar{z}yx\bar{b}\bar{a}$
   (d) $(\bar{y} + z)(y + z)$
   (e) $\bar{u} + \bar{v} + \bar{w} + \bar{x} + \bar{y} + \bar{z}$
   (f) $(\bar{z} + \bar{y} + \bar{x} + \bar{b} + \bar{a})(\bar{z} + \bar{y} + \bar{x} + \bar{b} + a)(\bar{z} + y + x + b + \bar{a})$

9. (a) $\bar{y}z + yz$
   (b) $(y + z)(\bar{y} + z)$
   (c) $S_2 S_3 S_4 S_5 S_6 S_7 S_8 S_9 S_{10} S_{11} S_{12} S_{13}$
   (d) $P_0 + P_3 + P_4 + P_5 + P_6 + P_7 + P_8 + P_9 + P_{10} + P_{11} + P_{14} + P_{15}$

10. $P_3 = \bar{w}\bar{x}yz$

Chapter 9

1. (a) z
   (b) $\bar{x}\bar{y}\bar{z} + x\bar{y}z + \bar{x}yz + xy\bar{z}$ (cannot be simplified)
   (c) $\bar{x} + \bar{y}$
   (d) $\bar{x} + \bar{y}$

2. (a) $xz + \bar{w}x\bar{y} + \bar{w}yz + w\bar{y}z + wy\bar{z}$
   (b) $x + \bar{y}$
   (c) $\bar{y}z + \bar{x}\bar{z} + wy\bar{z}$
   (d) $\bar{x}\bar{y} + w\bar{y}\bar{z} + \bar{w}\bar{y}z + \bar{w}\bar{x}z + wxy + xy\bar{z} + wy\bar{z}$ (some variations of the three-variable terms are possible)

3. (1a) z
   (1b) $(x + y + z)(x + \bar{y} + z)(\bar{x} + y + z)(\bar{x}\bar{y}\bar{z})$
   (1c) $\bar{x} + \bar{y}$
   (1d) $\bar{x} + \bar{y}$
   (2a) $(\bar{w} + y + z)(w + x + y)(w\bar{y}z)(\bar{w}x\bar{y}\bar{z})$
   (2b) $x + \bar{y}$
   (2c) $(\bar{y} + \bar{z})(\bar{x} + y + z)(w + \bar{x} + \bar{y})$
   (2d) $(w + x + \bar{y} + z)(w + \bar{x} + y + z)(w + \bar{x} + \bar{y} + \bar{z})(\bar{w} + x + \bar{y} + \bar{z})$
   $(\bar{w} + \bar{x} + y + \bar{z})$

4. (a) $\text{SDNF} = \text{SCNF} = \bar{x} + \bar{y} + \bar{z}$
   (b) $\text{SDNF} = \text{SCNF} = \bar{x} + \bar{y} + \bar{z}$
   (c) $\text{SDNF} = \text{SCNF} = \bar{x} + \bar{y} + \bar{z}$
   (d) $\text{SDNF} = \bar{x}\bar{z} + xz$
       $\text{SCNF} = (\bar{x} + z)(x + \bar{z})$
   (e) $\text{SDNF} = \bar{w} + \bar{y} + xz$
       $\text{SCNF} = (\bar{w} + x + \bar{y})(\bar{w} + \bar{y} + z)$

5. (a) $y + \bar{z}$
   (b) $w + x + y + z$ (cannot be simplified)
   (c) $x + y$

6. (a) $z + x\bar{y}$
   (b) I
   (c) $w\bar{y}$

7. (a) $x\bar{y} + vy + v\bar{w}z$
   (b) $p\bar{r}s + \bar{q}\bar{r}s + \bar{r}\bar{s}t = \bar{r}\bar{s}(p + \bar{q} + t)$

8. (a) $\bar{w}\bar{y}z + \bar{w}y\bar{z} + wyz + wx + x\bar{y}$
   or $\bar{w}\bar{y}z + \bar{w}y\bar{z} + wyz + wx + x\bar{z}$
   (b) $\bar{p}rs + p\bar{r}s + \bar{p}\bar{q}s + p\bar{q}\bar{r} + p\bar{q}s$ (five other possible answers, all using the first two terms)
   (c) $x\bar{z} + \bar{x}z + \bar{w}y + w\bar{y}$

9. (a) $(x + y + z)(\bar{w} + x + z)(\bar{w} + x + y)(w + \bar{y} + \bar{z})$
   (b) $(p + r + \bar{s})(\bar{p} + \bar{q} + \bar{r})(\bar{p} + \bar{r} + s)(\bar{q} + s)$
   (c) $(\bar{w} + \bar{x} + \bar{y} + \bar{z})(\bar{w} + x + \bar{y} + z)(w + \bar{x} + y + \bar{z})(w + x + y + z)$

10. Use conjunctive normal function, $(w + x + y)(w + \bar{x} + z)(\bar{w} + \bar{x} + y)$
    $(\bar{w} + x + z)$ seven NOR gates (including two as negaters)

11. $(\bar{w} + \bar{x})(w + y + z)(x + y + \bar{z})(w + x + z)$ eight NOR gates (including three as negaters)
    With 'don't care' conditions, conjunctive normal function simplifies to
    $(\bar{w} + \bar{x})(y + \bar{z})(w + y)(w + x + z)$
    and the disjunctive normal function to
    $\bar{w}xy + \bar{x}yz + w\bar{x}\bar{z}$
    Either of these two circuits also uses eight NOR gates.

12. (a) $(x\overline{\bar{x}\bar{y}z}) + (z\overline{\bar{x}\bar{y}z})$ (uses four NAND gates)
    (b) $(\overline{xwx\bar{y}z\,wxyz}) \vee (\overline{zwx\bar{y}z\,wxyz})$ (uses seven NAND gates, including two negaters)
    (c) $(y\overline{wx}) + (\bar{z}\overline{wx})$ (uses five NAND gates, including one negater)

Chapter 10

1, 2, 5, and 6.
   (a) $\bar{x}, z$
   (b) $\bar{x}\bar{y}, \bar{x}\bar{z}, \bar{y}z, y\bar{z}, xy, xz$
   (c) $xz, \bar{w}x\bar{y}, wxy, \bar{w}yz, w\bar{y}z$
   (d) $\bar{y}, \bar{w}x, x\bar{z}, w\bar{z}, \bar{w}z, \bar{x}z, w\bar{x}$
   (e) $wz, \bar{v}\bar{y}z, \bar{w}\bar{y}z, x\bar{y}z, wx\bar{y}, \bar{v}w\bar{y}, \bar{v}w\bar{x}y, \bar{v}\bar{w}\bar{x}z, vw\bar{x}y, v\bar{x}yz$

3 and 4.
   (a) $\bar{x} + z$
   (b) Several possibilities each containing three terms
   (c) $\bar{w}x\bar{y} + wxy + \bar{w}yz + w\bar{y}z$
   (d) $\bar{y} + \bar{w}x + w\bar{z} + \bar{x}z$
   (e) $wz + \bar{w}\bar{y}z + wx\bar{y} + \bar{v}\bar{y}z \ \vee\ \bar{v}\bar{w}\bar{x}y + vw\bar{x}y$ (other possible answers for
   the three-variable terms)

7. (a) No simplification of disjunctive normal form is possible (sixteen terms)
   (b) $z + x\bar{y} + v\bar{w}\bar{y}$
   (c) $y\bar{z} + v\bar{x}\bar{z} + w\bar{x}\bar{z}$

8. (a) Essential prime implicants give
$$\bar{w}\bar{z} + x\bar{y} + wz + \bar{x}y$$
   Non-redundant cover obtained by addition of any ONE of
$$\bar{w}\bar{x}, \bar{y}z, \bar{w}\bar{y}, \bar{x}z$$

   (b) There are twelve prime implicants,
$$\bar{w}\bar{x}\bar{y}, \bar{w}\bar{x}z, \bar{w}\bar{y}z, \bar{w}\bar{x}y,$$
$$\bar{w}\bar{x}\bar{z}, wx\bar{y}, wx\bar{z}, w\bar{y}\bar{z},$$
$$\bar{x}\bar{y}\bar{z}, \bar{x}yz, x\bar{y}z, xy\bar{z}$$
   Non-redundant cover can be obtained from several choices of six terms,
   e.g.
$$\bar{w}\bar{x}\bar{y} + \bar{w}\bar{x}z + w\bar{x}y + w\bar{y}\bar{z} + x\bar{y}z + xy\bar{z}$$

   (c) $\bar{v} + \bar{w} + \bar{x} + \bar{y} + \bar{z}$

9. $y + wx + \bar{v}w$

10. (a) $\bar{w}\bar{y} + \bar{y}z$
    (b) $\bar{w}\bar{y}z + \bar{w}xy + wx\bar{y} + wyz$
    (c) $\bar{x} + \bar{v} + wy$

Chapter 11

2. $$\begin{bmatrix} 1 & x & xy & (x+y) \\ x & 1 & y & (x+y) \\ xy & y & 1 & xy \\ (x+y) & (x+y) & xy & 1 \end{bmatrix}$$

3. $$\begin{bmatrix} 1 & xy \\ xy & 1 \end{bmatrix}$$

5. (i)

$$\text{L.H.S.} = \begin{bmatrix} 0 & 0 & 1 & 1 & 1 & 1 & 1 & 1 \\ 1 & 1 & 0 & 0 & 1 & 0 & 1 & 1 \\ 1 & 0 & 1 & 0 & 0 & 1 & 0 & 1 \\ 0 & 0 & 0 & 1 & 0 & 0 & 0 & 1 \end{bmatrix} = \text{R.H.S.}$$

(ii)

$$\text{L.H.S.} = \begin{bmatrix} 1 & 0 & 1 & 1 \\ 1 & 0 & 0 & 1 \\ 1 & 0 & 1 & 1 \\ 0 & 0 & 0 & 0 \end{bmatrix} = \text{R.H.S.}$$

(iii)

$$\text{L.H.S.} = \begin{bmatrix} 1 & 0 & 0 & 0 & 0 & 1 & 0 & 0 \\ 0 & 0 & 0 & 0 & 1 & 0 & 1 & 0 \\ 0 & 0 & 1 & 0 & 1 & 0 & 1 & 0 \\ 0 & 0 & 0 & 1 & 0 & 0 & 0 & 1 \end{bmatrix} = \text{R.H.S.}$$

(iv)

$$\text{L.H.S.} = \begin{bmatrix} 0 & 1 & 1 & 0 \\ 0 & 1 & 0 & 0 \\ 1 & 0 & 1 & 0 \\ 1 & 0 & 0 & 0 \\ 0 & 0 & 0 & 0 \\ 1 & 0 & 1 & 0 \\ 0 & 0 & 0 & 0 \\ 1 & 1 & 1 & 0 \end{bmatrix} = \text{R.H.S.}$$

Chapter 12

2. $x = \overline{b}$

3. p; $\overline{q}$; $p\overline{q}$; $p + \overline{q}$

4. The twelve solution pairs fall into two groups

a	b
0	$r \equiv s$
r	$r \equiv s$
s	$r \equiv s$
rs	$r \equiv s$
$r\overline{s}$	$r \equiv s$
$\overline{r}\overline{s}$	$r \equiv s$
$s \to r$	$r \equiv s$
$r \equiv s$	$r \equiv s$
r	$\overline{r}\overline{s}$
rs	$\overline{r}\overline{s}$
$s \to r$	$\overline{r}\overline{s}$
$r \equiv s$	$\overline{r}\overline{s}$

5. $b(a + c)$; $\overline{a}\overline{b}\overline{c} + b(a + c)$

6. No solutions for x and y in terms of a, b, and c and 64 solutions for a, b, and c in terms of x and y.

7. $a \equiv b$; I

9. $(\overline{a} + b)c$; $\overline{a} + bc$; $\overline{a}\overline{b} + bc$; $\overline{a}(b + c) + bc$

Chapter 14

1. (a)

		Input	
		0	1
Present state	0	1/0	3/0
	1	3/1	2/1
	2	0/0	0/0
	3	1/0	1/0

(b)

$$\mathbf{T}^0 = \begin{bmatrix} 0 & 1 & 0 & 0 \\ 0 & 0 & 0 & 1 \\ 1 & 0 & 0 & 0 \\ 0 & 1 & 0 & 0 \end{bmatrix} \quad \text{and} \quad \mathbf{T}^1 = \begin{bmatrix} 0 & 0 & 0 & 1 \\ 0 & 0 & 1 & 0 \\ 1 & 0 & 0 & 0 \\ 0 & 1 & 0 & 0 \end{bmatrix}$$

2. 0, 0, 0, 0, 0, 1

3. (a) z
   (b) x

4. The relationship is $(\overline{y \rightarrow x})\,(\overline{x \rightarrow y}) = 0$
   The constraint is either
         (i) $xy = 0$
   or (ii) $x + y = I$

5. $\begin{bmatrix} z' = z \\ y' = \bar{p}y + p\bar{x}\bar{y} \\ x' = py + \bar{p}x \\ w = py \end{bmatrix}$

   $\begin{bmatrix} z' = I \\ y' = \bar{z} + (p + y)(\bar{p} + x\bar{y} + \bar{x}y) \\ x' = p\bar{x}yz \\ w = p\bar{x}yz \end{bmatrix}$

6.

Present state	0 1 2 3 4 5 6 7	0 1 2 3 4 5 6 7
Input	0 0 0 0 0 0 0 0	1 1 1 1 1 1 1 1
Next state	6 2 7 3 2 2 3 3	6 2 7 3 0 0 1 1
Output	0 1 0 1 0 0 0 0	0 0 0 0 0 0 0 0

$$T^0 = \begin{bmatrix} 0 & 0 & 0 & 0 & 0 & 1 & 0 & 0 \\ 0 & 0 & 1 & 0 & 0 & 0 & 0 & 0 \\ 0 & 0 & 0 & 0 & 0 & 0 & 0 & 1 \\ 0 & 0 & 0 & 1 & 0 & 0 & 0 & 0 \\ 0 & 0 & 1 & 0 & 0 & 0 & 0 & 0 \\ 0 & 0 & 1 & 0 & 0 & 0 & 0 & 0 \\ 0 & 0 & 0 & 1 & 0 & 0 & 0 & 0 \\ 0 & 0 & 0 & 1 & 0 & 0 & 0 & 0 \end{bmatrix} \quad \text{and} \quad T^1 = \begin{bmatrix} 0 & 0 & 0 & 0 & 0 & 0 & 1 & 0 \\ 0 & 0 & 1 & 0 & 0 & 0 & 0 & 0 \\ 0 & 0 & 0 & 0 & 0 & 0 & 0 & 1 \\ 0 & 0 & 0 & 1 & 0 & 0 & 0 & 0 \\ 1 & 0 & 0 & 0 & 0 & 0 & 0 & 0 \\ 1 & 0 & 0 & 0 & 0 & 0 & 0 & 0 \\ 0 & 1 & 0 & 0 & 0 & 0 & 0 & 0 \\ 0 & 1 & 0 & 0 & 0 & 0 & 0 & 0 \end{bmatrix}$$

7. $\{1, 3, 0, 1\}$

Chapter 15

1. (a) 0 1 1 1 0 0 0 1 0 0 0 0 0 0
   (b) 0 1 0 1 0 0
   (c) 0 0 0 0 0 0 0 0 0 1 1 1
   (d) 0 0 0 1 1 1

2. (a) The cycle resulting from the disjunction is

0 0 0 1 1 1 1 1 1 1 1 0 0 0 1 1 1 1 1 0 1 1 1 0 1 1 1 1 1

and that from the conjunction is

0 0 0 0 0 1 0 0 0 1 0 0 0 0 0 1 1 1 0 0 0 0 0 0 0 0 0 1 1 1

(b) The cycle resulting from the disjunction is

0 0 0 1 1 1 1 1 1 1 1 1

and that resulting from the conjunction is

0 0 0 0 0 0 0 0 0 1 1 1

5. A run-in of 0 0 1 is followed by a cycle of

0 1 0 0 0 1 1

Index

INDEX

Matrices (*cont.*)
    switching   170–2
    terminal   172
    transition   250–1
Matrix, determinant of   182
    input–output   186
    null   176
    product   177
    terminal   172
    transition   250
    transpose   179
    unit   176
    universal   176
McClusky, E. J.   158
Mealy model   242–3
    conversion to   246–7
    state and output vectors of   251
    state diagram of   248–9
Minimization   107, 110, 155–6
    (see also simplification)
Modus ponens   82
Monostable   294
Moore model   243
    state and output vectors of   251
    state diagram of   250
Multiplier   234–8

NAND, see NOR/NAND
NBCD   21
Negater   98, 102, 105
Negation, graphical method   95, 96
    of a circuit   88
    of a proposition   62
    of a signal   98
Normal form   54
    conjunctive   54
    designation number of   129–32
    disjunctive   54
    NOR circuit from   106–10
    simplest   137
    special   55, 56, 73

NOR/NAND circuit, simplification of
    108–12, 148–50
    circuit design   106–12
    function   73, 75–7
    half-adders using   217–8
    gate   97, 98, 101, 105–15
    gates with two inputs   113–5
Number systems   1–28
    binary   1
    conversion of   4–13
    decimal   2
    octal   3
    sexadecimal   13

Operations, arithmetic   13
    cap   36
    complementation   35
    complete sets of   74–6
    conjunction   63
    cup   36
    difference   43
    disjunction   63
    implication   68, 76, 99
    intersection   35
    negation   62
    NOR, NAND   75–7
    switching   89
    symmetric difference   43
    union   35
OR, see AND/OR
Order relation   53
Output vector   251

Parity   23
Polish notation   118–24
    connectives for   119
    early-operator   120, 123
    forward   119
    late-operator   120, 122, 123
    reverse   123
Postfix operators   123
Predicate calculus   80